Disobliging Reality

Heckling the Sly Illusionist of the Here and Now

Frank Juszczyk, PhD

Frank Juszczyk is the author of a romantic adventure novel entitled, Our Gal Someday: A Scalar Signature Western

Archway Publishing books may be ordered through booksellers or by contacting:

Archway Publishing
1663 Liberty Drive
Bloomington, IN 47403
www.archwaypublishing.com
1 (888) 242-5904

ISBN: 978-1-4808-2615-1 (sc)
ISBN: 978-1-4808-2616-8 (e)

Library of Congress Control Number: 2015921364

Print information available on the last page.

Archway Publishing rev. date: 1/11/2016

This book is written for Jean, because of Jean, warm and sassy, always making divine mischief and lifting my heart to the nearest, most perfect paradise. No kismet was ever more welcome and even between the worlds now hovers so close, so close.

Preface

This book was written on the fly. My heart's companion, Jean, had made her passing less than two months before I began writing it. We had been together for forty-four years and married for forty.

Many of you know how inexpressibly overwhelming grief can be. You know the tumult of emotions that consume your days following such a profound dismemberment. Friends suggested that I begin writing to dissipate some of the intensity of my sorrow. I took their advice, not caring much what I was writing about or what I had to say. Like so many who endure such a loss, I was as angry as I was saddened. What was it now that separated me from the woman who was the sum of my totality? Why did everything in my life continue with business as usual when she was no longer a physical part of it? Why did reality remain the same without her in it? She had defined reality for me for a major part of my life. How could it persist without her participation in it? What was it about this reality that interrupted our joint being?

I became fixated on that specific question. I told myself that I would discover the operative dynamic that split us apart, that dislocated my place in this world and exposed the existence of another place where she had gone but that seemed impenetrable to me. Why had there to be a here and a not-here that were tied together by the same shared love, loss, joys, memories, thoughts, and values that inspired two lives lived as one?

My beloved, always an artist and musician, and I shared a passionate interest in quantum physics because of its incorporation of consciousness and paradox into its application. We were thrilled by this new dimension of scientific exploration that appeared to answer so many previously puzzling inconsistencies in what passed for ordinary reality. It was adventure and discovery on a level we could not have imagined before the 1970s. We ate it up. We read constantly and began attending Matrix Energetics seminars around the Northwest, West, and Southwest. Miracles happened. Minds were blown. We had come home.

Then home became a house divided. She moved, and I remained. I had overlooked a clause in our lease. There was a condition: nothing is permanent. After so much schooling in quantum physics, I couldn't help suspecting an illusion in play. So as I wrote to assuage my sorrow, I tried to uncover the nature of the illusion and how it works. This book is the result of my investigation.

I did not compose a rough draft and try deliberately to organize the material. I did very little editing and rewriting (not more than a sentence or two here and there). The words flowed, and I entered them into the computer. After all, I wasn't trying to write a book; I was trying to lessen the anguish I was feeling. Frequently, when I reread what I had written, it sounded strange and not at all like me. There were unusual patterns of thought and unfamiliar turns of phrase, as if the words were coming from somewhere other than the well-worn left hemisphere of my brain, or even not from me at all. I believe they were coming from an outside source. I believe my deep distress leapfrogged my awareness into another dimension of knowing. I don't exclude the possibility that Jean was a contributing author as well.

I hope you find the result useful and even entertaining. I still don't know why I remain here. Not everything can be contained in our specialized consciousness. The big stuff is just too big for our puny processing capacity. Yet we can think bigger than we usually do to some extent and continue beating the bush for that elusive hundredth monkey despite (and to spite) whatever our less-than-definitive, malleable mainstream science tells us is possible.

Introduction

What you are going to read (given the cautions alluded to in this introduction) is some complicated stuff about consciousness, reality, and yourself. If you are beginning without any previous general knowledge of quantum physics, you will probably be perplexed and disinclined to read beyond the first paragraph. I don't apologize for that. This is not a primer to quantum physics. What there is of it in the following pages, however, a practicing theoretical physicist would consider to be quantum physics at a pretty elementary level.

Even so, I hope I have not gone completely astray from the main concepts and principles of established quantum theory. At best, this is fringe or unconventional quantum physics in the way it relates to what we consider to be our familiar reality and how we experience it.

If you wish, as an average reader, to pursue what I am setting down, you will have quite a bit of catching up to do. If your curiosity is sufficiently piqued, I recommend that you read one or all of the following three books to get you started:

Quantum Enigma by Bruce Rosenblum and Fred Kuttner, Oxford University Press, 2006.

Bridging Science and Spirit by Norman Friedman, The Woodbridge Group, 1993.

Taking the Quantum Leap by Fred Alan Wolf, Harper & Row, 1989.

You will certainly find out if you have a taste for the quantum physics perspective on being alive in a world you are creating from moment to moment. Then, if you choose to go on, you can add supplemental reading along the way to fill in whatever gaps appear in your understanding. I know it sounds like some work, but the payoff will be worth it should you press on. As a result, whatever happens to you from now on in your life won't be so scary.

Also, some of your friends and relatives will find better things to do than stop by to see you anymore. That's because the surprising element of what I am presenting is that most people simply don't want to hear it. Even though the essential theories of quantum mechanics have been around for a hundred years, they're still the elephant in the room. No matter how chaotic and damaged people's lives have become, they would rather settle for a bogus external reality than face the prospect of being responsible for their own circumstances. That would mean never having the option of being a victim and never being able to blame others or operative causes, including fate, for getting the short end of the stick.

I'm figuring that if you already have enough independent spirit to risk your comfort zone, your deeper involvement in comprehending what I am presenting will make *Disobliging Reality* a field trip instead of a detached contemplation of ideas having little or no effect on the rest of your life. If you try it, doubtless you'll like it.

1

Reality is merely an illusion, albeit a very persistent one.
—Albert Einstein

Everyday or ordinary reality begins with a quantum wave function of infinite possibilities, one of which is collapsed into a single, specific, physical configuration as an object or thing. This process is initiated by either individual or group choice or both simultaneously. The catalyst for this transformation is the individual or group morphic field.

Morphic field is a term that was created by biologist Rupert Sheldrake to describe a field that organizes a characteristic structure and pattern of activity for both individual and collective members of the field. The term includes behavioral, social, cultural, and mental as well as morphogenetic fields. Often, the group morphic field directs the choices of the individual one. The individual morphic field, in turn, just as often feeds the group field with attention, emotion, and information, making the group field larger and stronger. The group field then creates what we know as consensus reality. This is all the physical stuff we see, hear, taste, smell, and touch and with which we associate strong feelings of attraction and aversion. This is what we take finally to be the real world.

Occasionally, something happens to one's individual morphic field that shakes one's trust and belief in what the group morphic field has created. This is a dramatically alarming and often traumatic turn of events. People who have invested deeply in the consensus reality fear what the loss of its foundational mind-set will bring about. For the most part, they fear that their belief in what is real will be jeopardized. In that case, their personal identity will be discredited as well. Therefore, maintaining the validity of both their individual morphic fields and of the consensus reality in which they have placed all their trust becomes an issue of basic survival.

The disheartening aspect of this state of affairs is that there are always more believers than doubters. The consensus reality is a consensus thanks to the majority of believers who nurture it. This makes life distressing for the minority of doubters, who live under the consensus reality's overbearing and imperious rule. The morphic field that creates consensus reality raises a screen between its compelling version of reality and what lies beyond. No alternate reality beyond the screen seems possible. But the doubters cannot be reclaimed by the believers. The captivating lie of consensus reality is too enormous to sustain itself. Inconsistencies appear, and contradictions surface. The believers ignore or suppress the anomalies, but the doubters are lured by their mysterious promise. If consensus reality is not as flat and predictable as it is incessantly shown to be, then unknown wonders may unfold beyond the tenuous limit of its prescribed yet arbitrary margins.

The doubter seizes the believer by his or her throat and demands, "Why this, but not that? What selective force of gravity bows us down to so scant a view of what can be? What turns us aside from our heart's true compass and screens us from a higher aspect of ourselves?" It is the reality that we know and foolishly promote as our binding environment, our familiar and reassuring stretch of the imagination. What we are, we are, and no more than what we are is or can be. There is a stink to its finality. First we settle; then we rot. We wander in a maze of our own making, lost, but with a hope that we are getting somewhere. These are merely circumstances we have chosen to honor. They are the rules we have selected to play by. Ignore them, and the game is over. Circumstances lose their relevance. The screen collapses, and what is separate unites with itself in a bliss of totality.

But our continual reliance on our metaphorical sorting tray fore-stalls this immense revelation. Raw experience must be sorted into the appropriate shallow, cuplike depressions in the bottom of the tray. Without the sorting tray, reality loses its conventional meaning, which is the steering mechanism for our personal relevance. The things created by our individual and group morphic fields are sorted into their appropriate cups first by separating them one from another out of the indistinct wave function of all possibilities. Then they are identified by their apparent differences, and assigned names or labels so that their exclusive isolation may be conveniently recognized over and over again without having to repeat the original sorting process. This process includes everything that appears in our reality, from soup spoons to space shuttles. Reality is made automatic, confusion is avoided, and there are no unpleasant surprises. Reality is comfortably humdrum, resting on the bottom of the sorting tray.

The doubters are outraged. So much has been lost, set aside, dismissed, ignored, rejected, excluded, discarded, and disallowed simply because the consensus majority can't "get its head around it." Look at the unrecognized sorting process in that defensive phrase. Why does the head have to be the defining agent for understanding something? Is consciousness so restricted that it functions only above the neck? And then how does one get one's head "around" something? Apparently it is a confining enclosure. If the head can't contain it, then whatever it is isn't worth the effort to put it in there in the first place. The head becomes the sole arbiter of what is real.

Then there is the "I" who can't get its head to go around whatever the head needs to contain. Where is this "I"? It sounds as though it is separate from the head, perhaps able to recognize the head's limits and spare the head unnecessary aspirations toward enlarging its capacity. The "I" sounds as though it is more discerning than its head, thus relegating its head to a secondary function, like tweezers or a pair of pliers, a tool used for grasping something. So the "I" confesses its inability to "get its head" to work properly. If it could, the "I" would gain greater knowledge and understanding. But alas, the "I" can't always get its head to do what it wants done. So the "I" goes about in a state of torpid stupidity, the victim of its malfunctioning head, and we see just how much the consensus "I's," the creators of our reality, are actually capable of doing under such adverse circumstances. We, the doubters, are left in complete ignorance of what the creators of our reality can't get their heads around (because they are incapable of acknowledging its existence). It is infinitely greater and more deeply sustaining than what they can.

So how can we insinuate ourselves into a richer, deeper, multilayered, multidimensional flow of the reality that lies beyond the consensus screen? First, don't identify with the prevailing version of reality. Don't legitimize or validate it with your participatory attention in the form of anticipation, expectation, speculation, trepidation, or imagined outcomes as a formulated destination. In other words, don't let the left hemisphere of your brain call the shots, as it has for most of your life.

For a helpful perspective on just how your left brain creates your reality, read Jill Bolte Taylor's fascinating account of what it's like to live without it in *My Stroke of Insight*. Paradoxically, Taylor is forced to describe her transformation only in left-brain terms as, first, she is a scientist (a neuroanatomist) and must proceed in a logical, analytical fashion, which is a distinctively left-brain process, and second, verbalization itself in the use of language is also exclusively a left-brain activity. In other words, Taylor must remain on this side of the reality screen in order to communicate what happens on the other, nonlogical, nonverbal side.

Is there no escape from this linear and sequential tyrant? Not, I'm afraid, within the perceived limitations of this reality. But there is a loophole. It is the fact that what we experience as reality is both real and not real at the same time. This means we have the option to choose. Albert Einstein said, "Reality is merely an illusion, albeit a persistent one." It is the persistence of reality that ensnares us more than the assumed reality itself.

Reality nags. It tugs at our clothing like an insistent child. It is obnoxious and annoying. It wants our exclusive attention. It won't go away. It is completely self-absorbed and preoccupied with its own self-importance. If there is anything beyond itself, it is too busy trying to satisfy its own needs to perceive what that might be. We try to elude it in various ways, to make it go away using obsessive distractions, consciousness-altering substances, or mind-numbing spiritual practices. Yet when we dare to peek out from under our protective blankets, it is always still there. Sometimes—as in the case of sudden, overwhelming loss—it will yank our blankets away, leaving us exposed and helpless before it. This is the in-your-face version of reality that we dread the most.

So the most immediate question is, "How do we reduce or eliminate the *persistence* of consensus morphic field reality?" More to the point, let's phrase the question in an open-ended way: "If I were to reduce or eliminate the persistence of reality, what would that be like?" Where does our open-ended question take us? Not to our imaginations. That's just free-associating our past experiences. Not to our speculative reasoning. That too depends upon our past experiences but in a deliberative, selective fashion. In either case, we resort to searching our left brain's archives for relevant past experiences. We search our existent reality for something with which to construct a reality we have not yet experienced. Obviously, our open-ended question must take us somewhere we've never been.

How do we get there? Maybe we're already there but can't recognize it because of the compelling nature of where we've been all this time. All of our "somewheres" have been created for us by the reality we want to diminish or delete. Let's begin with a supposition already beyond our reality's reach: reality *can't* be real. What if the reality we think we know is not merely screening out a greater reality we don't know but is itself unreal to begin with? More and more physicists are supporting this conclusion since Einstein first proposed it.

The issue is not whether reality is an illusion—remember, it *can't* be real—but rather, what makes this illusion so damned persistent? The *how* of reducing reality's persistence depends upon discovering *what* makes it so persistent. The most persuasive aspect of its persistence is that, basically, it is unpredictably *dangerous*. This quality gives reality a random potential that defines our relationship to it. Reality becomes the loose cannon of our daily existence. We try to secure it in various ways, but it always seems to break loose when we least expect it and threaten our safety with its uncontrollable and haphazard dislocations. We sense this negative potential in reality and resonate with it at a very deep level. Given this latent power over us, reality romances us with both its allure and its sinister ambiguity. Will it respond to our tender advances or destroy us with callous indifference? We are captivated by the dramatic possibilities of its dualistic potential, so it intrudes on our awareness relentlessly.

Yet it *can't* be real. Countless studies (such as the following representative examples: "Looks Can Deceive: Why Perception and Reality Don't Always Match Up" by Christof Koch in *Scientific American*, July/August 2010; "The Illusion of Reality: A Review and Integration of Psychological Research on Hallucinations" by R. P. Bentall in *Psychological Bulletin,* Vol. 107(1), Jan. 1990, 82-95; "Sensory Illusions" by Debra Speert PhD in *Society for Neuro-Science,* BrainFacts.org) have revealed to us the deceptive character of our senses, our inclination to draw false conclusions from plain evidence and to mistake wishful thinking for certain knowledge. We all have our agendas, and we shape our experiences to suit their requirements. The reality that scares and intrigues us is the same reality that vanishes when we sleep, but we reinvent it every time our waking consciousness returns. We are convinced that reality is continuous, that, unlike us, it never sleeps, but we reinvent our proofs of that along with the reality itself, as if the proofs had some existence independent of the reality that creates them.

For hundreds of years now, we have been reverse-engineering our universe, our reality, to find out how it works so as to manage it properly and reduce its dangerous propensities. However, all that reveals itself over and over is our own consciousness staring back at us. The double-slit experiment, which proved the simultaneously material and immaterial nature of reality, has led to the double-blind method of verifying reality in the hope of tricking our consciousness into forgetting what it does, which is to collapse an infinite range of possibilities into a single, specific thing or event and then call it an independently existing reality. We play with shadows and never realize that we are casting them.

It *can't* be real. Consider the Goldilocks enigma. Why are the physical "laws" of the universe just right for developing and sustaining life? Very slight changes in the balance of variables that go into constructing our reality would bring about our doom, not to mention the elimination of any likelihood that we ever would have appeared here at all. According to Peter Woit, in his review of Paul Davies' book, *The Goldilocks Enigma: Why is the Universe Just Right for Life?*, the most recent research in theoretical physics suggests that our universe is only one of an "unimaginably large number of possibilities" and that "if one can get our observed universe this way, one can also get just about any variation of it. . . " (1).

So our universe (our observed reality) *can't* be real in terms of an exclusive right to be as it is. There are competitive options (possible parallel universes) that have the same degree of reality as ours, which raises the question of how our reality is determined in the first place when you can have so many different versions of it. In the end, we are left with the unsettling intimation that our reality, both personal and cosmological, *can't* be real unless we give it our conscious (or even unconscious) permission to be so.

2

Switching On and Off

Now let's look at microgeny. This refers to the pulsating nature of reality. Now you see it, now you don't. The frustrating aspect of it is that you can't see it when you don't. That is, you can't see yourself not seeing it, so all you see is the seeing of it. It persists in the same way the persistence of vision functions. Think of the old cartoon flip books. A cartoon figure is drawn in a series of static poses on multiple pages. The poses are depicted as separate stages in a process of movement. When the pages are riffled or flipped quickly in sequential order, the cartoon figure appears to be in continuous motion because our persistence of vision selects the cartoon figure for its attention and ignores the intervals between the pages, giving the cartoon figure a continuous presence uninterrupted by the gaps in which it disappears. That is very similar to the way in which our consciousness perceives our physical reality.

Norman Friedman, in his book *Bridging Science and Spirit*, quotes physicist David Bohm on our flip-book reality: "The implicate order can be thought of as a ground beyond time, a totality out of which each moment is projected into the explicate order. For every moment that is projected out into the explicate there would be another movement in which that moment would be injected or 'introjected' back into the implicate order " (137).

This pulsation has been called by Bohm "enfolding" and "unfolding" or "involution" and "evolution." Because the pulsation is consistent, we miss the interruptions in the overall process. The unfolding persists while the enfolding disappears. During the enfolding phase, our reality exists in a completely different system of reality, which could contain many different universes. Could we handle that perceptive experience? Could we get our heads around it? Ultimately, can we handle the truth?

Perhaps another level of fear maintains the persistence of our reality, our fear of experiencing another reality greater and more complex than the one we know. Could it be the fear of finding ourselves to be greater than our self-created reality has taught us we can be? I think that's part of it. We fear the dangerous potential of the reality with which we are familiar because it contains unknown aspects that can hurt us, and we fear a completely different reality because its entire nature is unknown to us and is therefore potentially dangerous in all of its aspects. We can accommodate ourselves to the fears we know and share with others in a familiar reality that is limited by what we think we know, but we are overcome by the fearful prospect of confronting an alien reality that exists beyond our very capabilities of knowing.

But if we can disoblige the reality we know—that is, inconvenience it by no longer deferring to it and interrupt its unfolding phase so that its enfolding phase becomes apparent—we can enrich our ordinary reality with further dimensions of potential experience now hidden behind that reality's repetitive persistence. We can remove the screen of consensus reality so we can see and experience what lies beyond.

Interestingly, the Taoist art of *Taijiquan* was developed specifically to bring the pulsating nature of reality into our conscious and subliminal awareness. The structure of *Taiji*'s movements mimics the pulsating action of the universe. But instead of enfolding and unfolding or involution and evolution phases of movement, *Taiji* is predicated upon the oscillation of yin and yang.

In *Taiji* terms, involution corresponds to storing and evolution corresponds to emitting. The same correspondence applies to what *Taiji* masters consider the insubstantial followed by the substantial. It is yin followed by yang expressions of movement. This is the basis for the claim that when one attacks a master of *Taiji*, one is in effect attacking the power of the entire universe.

Unfortunately, *Taiji* is seldom if ever taught in this context in Western culture. It is usually offered as an esoteric mind and body discipline of therapeutic value, ignoring the fact that there is no distinction made between mind and body during the enfolding or yin phase of *Taiji* practice. Dualism is a yang characteristic, which is created out of the implicate yin totality of the universe. We are, in effect, this pulsating oscillation but are constituted in such a way that we are inclined to overlook our yin component. For us, especially as Westerners, it is yang, yang, yang throughout our reality. Asians have their own peculiar obsessions, but disavowing the yin component of reality isn't one of them. For us, it is always this instead of that.

Is there a way—or more properly, a no-way—that we can pay more attention to that and less to this? We are compelled by the circumstances of this reality to begin in a practical way, giving our yang-sodden left brains due recognition for their untiring persistence. That's what I've been doing in this discursive preamble. I've been explaining in left-brain language why left-brain reality is so obtuse and ultimately just flat wrong. You see, according to the latest quantum mechanics theory, three-dimensional reality is a holographic projection of encoded two-dimensional information, making it flat at its source and to us doubters, wrong in its oversimplified presentation of an inadequate and cheesy reality. And because it is inadequate for the scope of our curiosity and imagination, it must inevitably change to suit the expanding capacity of our collective awareness.

Former flat-earth believers are now sending probes to Mars. Their change in perception of what is possible brought about the corresponding change in the nature of their reality. A change in meaning is a change in being. According to David Bohm, "A change of meaning is a change of being. If we say consciousness is [the] content [of being], therefore consciousness is meaning. We could widen this to a more general kind of meaning that may be the essence of all matter as meaning" (qtd. in Friedman 81). If matter is equivalent to meaning, then matter must conform to the meaning that is consciousness. A flat earth becomes spherical when our consciousness of it alters its configuration. So by explaining conceptually the ways in which our consciousness creates our material reality, I am preparing our consensus consciousness reality for a dramatic paradigm shift.

We undermine our familiar logical, linear conception of reality first by presenting a logical, linear case against it. The left brain then entertains this new reorientation as a legitimate possibility, but only on a rational, speculative level. That, in itself, is not enough to cause a genuinely transformative change. To do that, we must connect with the implicate order of things, the larger reality beyond the screen of our ordinary perception. Our predisposition to do this depends upon the strength of our rational evidence for an alternate version of reality. A functional certainty (our everyday reality) is expanded to include a larger theoretical possibility (that our consciousness creates our reality), so we are not as resistant to the idea as we might have been had it been thrown at us without warning.

Considering the magnitude of the dynamic with which we are working—the pulsating nature of ultimate reality itself—we cannot bring about a change we can experience directly by merely winning a debate. Intellectual assent is not the same as a change of heart. We are seeking a new feeling state beyond conceptual understanding. That's the point where involution turns into evolution, where the unsubstantial becomes substantial, and we can see in two directions at once, thus allowing us to experience the pulsation of reality in the gap between the implicate and explicate orders of being. It's like becoming the pause between inhalation and exhalation, the breath of no-breath. It is "uncollapsing" wave function back into its infinite possibilities before an observation, in its obsessive persistence, forces some three-dimensional thing down our throats.

3

Double-Slit Awareness

What we need is a bilateral awareness, something like the electron interference screen used in the famous double-slit experiment, something that will reveal both the particle and wave aspects of reality. Our ordinary awareness is notoriously localized, so the addition of a complementary non-localized awareness would give us access to both sides of the reality screen. Let's refresh our basic understanding of how the double-slit experiment was set up and what it revealed.

The first double-slit experiment was devised by Thomas Young in 1801, using a beam of light that passed through two vertical slits scratched onto a soot-covered piece of glass. The light formed a pattern on a second screen behind it. Young discovered that the light passing through the two slits created an interference pattern, which could not occur if the light was made up of particles only. The result showed that the light was actually taking the form of waves.

In the 1960s, physicist Richard Feynman carried Young's experiment considerably farther. He created a thought experiment using electrons instead of photons. In the experiment, a gun shoots electrons at a wall with two tiny slits that can be open or closed. After passing through the slits, the electrons hit a detector. Passing through only one slit, Feynman predicted, the electrons would behave as particles, but if they passed through both slits, they would behave as waves. This was later proven to be true in subsequent physical experiments, showing that electrons could behave either as particles or waves. When behaving as waves, the electrons produced an interference pattern, but when behaving as particles, they produced a pattern of separate bands.

Ultimately, this implied a dual, unfixed, and fluid nature to reality. But even more astounding was the result of refining the experiment to observe which of the two slits a particle would pass through. What happened was that the act of observing which slit an electron would go through caused it to go through only one slit, creating the particle band pattern, but when no observation was made, the electrons passed through both slits simultaneously, creating the wave interference pattern. Whether the electrons behaved as particles or waves depended upon whether they were being observed or not. That is, observation determined whether the electrons behaved as physical particles or as nonphysical wave potential. Observation of the electron collapsed its wave function into a discrete particle, making the unsubstantial substantial, the enfolded unfolded, and infinite potential a physical reality. That's quite a payoff for just looking at something, but we do it easily, casually, at every moment of our lives. The process is that persistent.

Consider your awareness to be a double-slit experiment. Most of the time, it allows only particles to pass through one or the other of its slits. Occasionally—and this is rare—it allows wave function to pass through both slits at once. This is because you failed to observe the process for a moment. You forgot to look for the particle, and no particle was collapsed out of its preexisting wave function potential. There was a glitch in the matrix of your ordinary reality. But this sort of glitch is so rare and unexpected that you failed to notice your habitual particle creation and that instead of a particle, a wave appeared. Thus, there was a paranormal anomaly in your customary world-making, which, being paranormal, did not register in your awareness. So your double-slit experiment is continuing to prove that your world is made only of particles.

But having read my compelling explanation of why this cannot be, you are beginning to entertain the notion that you could be missing something, although nothing really convincing yet. But even preposterous suggestions can have some teasing effect on one's interest.

Remember that it takes only one exception to disprove a theory, and one wave appearing where there should be only particles disproves the theory that everything is made out of particles.

The important thing is that your awareness, used experimentally as a device for distinguishing a particle or material reality from a wave or nonmaterial reality, can free you from perceptual enslavement by a persistent illusion. Remember that your double-slit awareness can show you the dual nature of reality—its illusory physical aspect and its true wave potential, nonphysical aspect—so long as you monitor your observation of it. If your attention is fixed on observing particles passing through a single slit in your awareness, that is all you will see, but if your attention is unfixed and freely indeterminate—that is, uncommitted to a specific task—you will begin to see waves passing through both slits in your awareness.

First, you begin to disoblige your particle reality by destabilizing it. You make things inconvenient for it by doubting its legitimacy as the sole reality available. You stop letting it take you for granted as a blind adherent. You are no longer its fool. Its exclusive hold on your attention becomes more and more intermittent as your consensus reality mind-set wavers. At this point you are still operating on an ideational level. Ordinary reality remains as persistent as ever, and life goes on.

But think of it as a suspicion that your spouse or loved one is cheating on you. He or she may continue to look and behave in the normal way, but you notice a difference in the way you feel toward him or her. You look for signs of deceit, inconsistencies in behavior, or insincerity in expressions of loyalty, dependability, or devotion. You don't know if you can trust what you see and hear anymore. You develop a perceptual sensitivity you never had before. This conceptual alteration in your relationship to reality begins to involve specific feelings and becomes progressively less conceptual and more emotional. It creates a certain tone or mood that pervades your interaction with a reality you once trusted.

Typically, when there is a betrayal of trust within a romantic relationship, strong emotions of jealousy, anger, disillusionment, and possibly vindictiveness are aroused. In the case of a cheating reality, however, the injured party cannot so easily assign blame. After all, the disillusioned partner is equally responsible for the deceit by, albeit unknowingly, helping to create it. What emerges is a kind of pervasive, unconditional doubt and a permanent inclination never to allow one's trust to be so easily captivated again. A wry skepticism becomes the dominant mood involving all interaction with our sly illusionist of the here and now. Suspecting that you are not actually where you've assumed yourself previously to be, at least in any final sense, you can begin to disoblige the hard edges of an intrusive simulation.

You may marvel at its breathtaking beauty, its stupendous scope and complexity, its shocking horror, its comforting blandishments and glib indifference, but despite whatever it shows you, you sense that it requires your own complicity of attitude and feeling to give it density and dimension. Without your searching observation, your expectation of confirming the appearance of a particle, there are only waves of possibilities. Whatever you are seeing or experiencing could have been something else just as well. Uncertainty is the defining temper of your revised and expanded reality.

Cultivate your doubt and uncertainty as you would a tender seedling. Keep concern for its well-being an ever-present awareness in the back of your mind. Whenever you realize that reality is taking you for granted, disoblige it by recalling the presence of your seedling uncertainty. Assess its health and vitality. How is it doing? Does it seem strong and invigorated?

Above all, don't let it wither from neglect. It will eventually become the background to your awareness generally, the texture of your sense of presence in the world. The sly illusionist of the here and now will find it harder and harder to misdirect your attention from how the trick is done.

4

Wow, That Was Fast!

Continuing your double-slit experiment, you now know what the slits in your awareness can do. They are portals into different realities. By learning how to disoblige reality, you are creating a choice between experiencing a fixed, single-slit reality or a double-slit potential one. Now here's where things take a curious turn. There is a parallel version of this experiment that has become widely known and extensively applied, though its success rate is unknown because it was never conducted under strict scientific protocols. The similarity became apparent to me only after I had been considering the larger implications of the double-slit experiment.

Back in the late '60s and early '70s, like so many others in the twenty-something age range, I read Carlos Castaneda's books. They were exotic and strange at the time and promised a way of entering a transcendent reality beyond the mundane one we knew. They were full of secret knowledge developed from ancient times by an indigenous culture wiser than our own in the ways of spiritual transformation. In the late '60s, this kind of mind-altering discovery was gold.

Not only that, it involved using hallucinogenic plants to make it work. How could the search for one's inner divinity get any better? There were a lot of stoners out there who could hardly wait to experience being a luminous egg. Unfortunately, there was a lot more to becoming a Yaqui sorcerer than just getting totally ripped. Ideally, you had to get down to the Sonoran desert of Mexico and then find an authentic sorcerer who would agree to make you a brother practitioner.

But where did you go to sign up? I imagine many dreams were con-
founded and even lives lost in that latter-day gold rush.

There was considerable controversy over whether Castaneda's
anthropological documentary was authentic research or just sto-
ry-telling. Even today, there are both devoted followers and unfor-
giving skeptics of Castaneda's work. It occurs to me, however, that
whatever Castaneda was up to, he may have been way ahead of the
curve when it came to intuiting quantum mechanics' principles in the
teachings of Don Juan.

Consider two major concepts that establish the foundation of Don
Juan's worldview—the *tonal* and the *nagual*. The *tonal* corresponds
to ordinary or consensus reality and the teachings for the right side
of man, represented by the reality we have chosen to disoblige. The
nagual corresponds to the interference pattern of wave potential
representing a separate reality of infinite possibilities and the teachings for the left side of man. It is to gain passage into this separate
reality that we disoblige our ordinary one. The solid mind-set of the
tonal must be thinned and its active intensity diffused to allow for the
revelation of the indeterminate mind of the *nagual*.

For the sorcerer Don Juan, the world is made of fields of energy
rather than solid objects. This view is true for the quantum physicist as well. The field of fields is the *nagual* itself. Victor Sanchez,
in *The Teachings of Don Carlos*, a handbook for the application of
Castaneda's Yaqui lore, noted that Don Juan distinguished the two
realities as being one in which you talk (the *tonal*) and one in which,
as a sorcerer, you simply act (the *nagual*) (15-17).

Acting is different from reacting, which is what most of us do. This
is a critical point. There is no calculation to acting, no implementation,
no assessment of outcomes, no modifying context at all. Acting within
the *nagual* is not, strictly speaking, spontaneous because it has a
source. It is not the random impulse of a scattered or immature men-
tality. Acting is being present, not just in every moment but in every
pre-moment. It is tapping into the timeless domain. Obviously, acting
takes place outside of a cause-and-effect dynamic. It is non-dualistic
because it originates outside of the play of polarities. Its source is

the totality of all that is. Sounds impressive as hell, but what does it mean?

Kim Russo, "The Happy Medium," appears in a television series called *The Haunting of …* in which she guides various celebrities through recapitulations of their paranormal experiences so they can better understand what the experience meant and why it came about in the first place. From time to time, she encounters subjects who she thinks show psychic abilities of their own. To discover the presence of such abilities, she will administer a test.

In one episode involving a male actor, he and Kim were standing in front of the closed door of a room in the house in which the actor experienced a paranormal event. Kim had the actor face the door and told him to identify ten items that were in the room. She told him not to think about it, try to imagine what might be in the room, or even hesitate before telling her. He rapidly listed six or eight items, and then they entered the room. The items he identified were all there. *That's* acting. Cause and effect are simultaneous. The information is already there, and the actor knew it. Reporting it to Kim was what introduced a time factor and the semblance of an outcome. When the actor knew what was in the room, he was in a different world, a separate reality. When he began to tell Kim about it, he was back in this world of physical objects and time.

It can be useful to test yourself in similar ways to find out just how dependent you are upon the limitations of consensus reality. Remember that you are not playing a guessing game. Nothing is hidden from you. There is nothing to figure out. Just *act*. It precedes your most immediate impulse. It already knows what you are about to do. It already exists as a probability before you make it known as an event in space-time. Get used to trusting it. It is even more you than you think yourself to be. Become cognizant of moving through a reality in which you are always lagging behind. In consensus reality you are continually catching up. In effect, you are habitually saying, "What happened? I only just got here."

In quantum physics terms, elucidated by Friedman in *Bridging Science and Spirit*, nature can bypass the law of the conservation of energy if the time period is short enough. In other words, the less time involved, the more energy is produced, enough energy to distort space-time and produce a quantum foam consisting of wormholes and bridges (black holes and white holes) that emit information into our universe (203-204).

Go back to Kim Russo's test for psychic ability. The shorter the time interval measured in any precise way, the greater the energy concentrated in diminishing areas of space. This space could be a sea of virtual quantum black holes blinking in and out of existence at an unimaginable rate (frequency). The holes are entry and exit points into and out of our universe. Black holes are exits, and white holes are entry points. They are called singularities. To create one of these singularities, such as knowing what is on the other side of a closed door or "seeing" through apparently solid matter, you must approach a timeless state by simply acting in the instant. Any hesitation dissipates energy, and the more energy is dissipated, the less space-time is distorted and the fewer the singularities produced. As a result, less information is emitted into our universe, including what is behind a closed door we can't see through.

Because consciousness pulsates in the same manner as the black and white holes that constitute the quantum foam, it is subject to the same time compression as they are—the less time involved, the greater the energy created. The bigger the job, the shorter the interval needed to "git 'er done." Think of those reports of people performing impossible tasks in emergency situations like lifting a car off of others pinned underneath. Obviously, your best option is to live as much as possible in a near-timeless state. This has nothing to do with being in a hurry. "Hurry" is relative to a standard of time that does not exist.

This may help. Einstein stated more than once that time is not a condition of our everyday reality. His point was that time has no real existence. It is just made up. Summarizing the growing viewpoint among free-thinking quantum physicists, Friedman, again in *Bridging Science and Spirit*, says "that space and time are constructs of the mind of the observer rather than inherent in any absolute way in the fabric of reality" (182). Yet time is the atomic "wedgie" for those of us living in a holographic reality. There is either too much of it or too little; it moves too slowly or too fast; it brings things into existence, and it takes them away. We save it, waste it, find it, lose it, make it, bide it, and kill it. But we never really know what it is because it isn't an "it." We wrestle with a ghost that leaves us with our underwear over our heads.

In our experience, time is an abstraction, a quality apart from an object even though an object may appear to be changed by it over time. Remember microgeny, the pulsation of reality? What appears to change an object is its weakening of energy after innumerable appearances and disappearances in our particular dimension. Reality is not continuous, so time is not a constant, and Einstein proved that time is relative in any case. No object exists long enough to change with age. It just becomes a less-perfect duplicate of what it was in all of its previous appearances. And remember that this process of seeming degeneration has no temporal context in itself at all. It is what it is, and we supply the time frame out of our own observation.

Consider this: when our observation of all the possibilities that exist as a wave function collapses one of them into a specific object, that object appears with its own complete history. A rock, let's say, that you observe on the ground is not only one of the infinite number of probable other rocks it could be in a field of infinite possibilities; if you examine the rock and subject it to tests to determine its age and discover that it is sixty million years old, it possessed that age the moment you created it with your observation of it. When you brought it into physical existence, you created its past as well.

It is this bogus past that we assign to objects, events, and even ourselves created by our own observation of them that haunts our everyday reality. It makes things appear and feel more authentically real than they actually are. It makes reality oppressively pervasive in every aspect of our lives. It creates cause and effect, timelines, trends, tendencies, life expectancies, and even its own disappearance with the "end time" of apocalyptic anticipation. It creates a colossal network of implications that can ensnare and terrify us like a heavy net.

But what if you stopped assigning a past, and therefore a future, to everything you observe? There would be no temporal context for anything. Everything would exist for you in the moment of observation as what it is in itself without any other significance to engage your speculative thought processes, your desires, your fears, your likes and dislikes. Talk about disobliging reality! You would feel as though you were walking through an artificial setting created from moment to moment as you pass by.

That's the feeling you want to install in your awareness. No, you won't lose your ability to function in the previous reality. It is far too persistent. But you will have a choice of which reality deserves more of your attention. This is one of the major benefits of *acting*. The less you clutter your consciousness with thought-constrained implementation and the continuity of things, both behind and ahead, the more you will disoblige consensus reality and see what is beyond its screen. Release your fresh and immediate perception from the withering necessity of progression. An interruption of continuity dissipates reality's persistence, and *acting* confounds the measurement that simulates time. You are deconstructing a point of view that is selective and therefore dualistic. In dualism there is conflict, and in conflict there is drama. Bring down the curtain on the play of perception.

Both *acting* and eliminating the historicity of reality opens both slits in your double-slit awareness. You have access to the interference pattern of wave potential because you are acting too fast for your own measurement. The singularities created by *acting* produce an effect similar to what physicist Fred Alan Wolf, in his book, *Mind into Matter*, refers to as tiny gaps in space-time "where forces are beyond comprehension" (qtd. in Friedman 204). You have shut down your single-slit particle counter and have entered double-slit super-space. You have eliminated measurement as an option by allowing a gap in space-time where cause and effect cease to function.

This measurement aspect of consciousness has caused no end of trouble for proponents of a totally physical reality. Wolf talks about what he calls a sudden "pop of reality" when the probability cloud of a single atom (which includes all of the possible positions of the atom) evaporates, leaving a single atom because a conscious mind noted a "real" measurement of the atom's location. From this perplexing phenomenon came Werner Heisenberg's uncertainty principle and Niels Bohr's complementarity principle, which in essence say that events cannot be connected in terms of cause and effect and the physical universe can never be known independently of what the observer chooses to observe.

Put another way, observer and observed are entangled on the quantum level once the intention for observing arises. The intent to observe causes the entanglement to begin. This means that if events *were* connected by cause and effect, the observer would be creating or causing what is observed. The observer and the observed are aspects of the same process. The double-slit experiment does not exist independently of the one conducting the experiment. Our reality is made up of our observation of it.

So the challenge for us doubters of an independently existing reality is making the observer redundant—making it an unnecessary accessory to an already adequately functioning consciousness. If the observer is regarded as indispensable, the impression is created that no other options exist than what the observer permits, and the infinite possibilities of a separate reality never enter our consciousness. The observer is necessary only to make real-looking stuff out of wave potential. And this stuff is awkward and clumsy—all that weight and mass and inertia. You get used to it, and then you think you can't do anything without it. You're moving furniture that you create by observing it. But you have the choice to impair the function of the relentless mechanism that makes it. The difference between struggling to move the furniture and getting rid of the furniture-maker who is creating the struggle is an enormous one.

To move furniture, you need to be in a cause-effect, space-time reality. To impair the process that creates the illusion of furniture, you need to enter a feeling state created out of a separate reality. Because your consciousness pulsates at a certain frequency, you have a default setting that creates real-looking holograms like popcorn at a carnival. So you have to go back to your two-slit awareness, open both slits, and put blinders on your observer. By creating a certain kind of feeling state, you can allow things to happen that are not under your conscious control, and your observer has trouble reproducing an everyday reality context for what is going on. You blur the usually sharply defined separation between the infinite possibilities of pure wave function and a specific, localized holographic form existing in space-time. Because both of your awareness slits are open, your observer can't focus its attention on both of the slits and what is passing through them simultaneously. Your observer cannot track both the speed and position of what is taking place and is left with no certainty at all. In this state, anything can happen. You can reconfigure habitual patterns of consciousness or of patterns of illness. You can visit other worlds and other realities and access data that was hidden from you previously. You can alter time and compress or extend space. You can create synchronicities in everyday experience. The choices are infinite.

You do not have to enter a hypnotic state, but you need to develop an abstracted one. The word *abstract* has meanings like, "to draw away the attention of," "to remove or separate," and especially for our purposes, "expressing a quality apart from an object." The abstracted state is not inherent in a particular thing. We are overriding particular things. That's what happens when we get dreamy or begin wool-gathering, and our attention slips away from our immediate physical surroundings. We get woozy, dopey, and inattentive and fall into a careless reverie. This is not a meditative state. Meditation is too outcome directed. People meditate for a reason. Meditation is a form of therapy. Abstraction doesn't have enough focus to be purposeful. You aren't engineering a result. You are allowing space for possibilities. Maybe this sounds like hair-splitting, but you don't need a mat and contorted positions to work some magic.

Find the feeling, but don't isolate it in your head like a thought. Allow this feeling to flow through your whole body. Specifically, if you are standing, let the feeling slide down your back and legs and into your heels. You will feel a drawing or pulling sensation in the middle of your back below your shoulder blades, which will make you feel as though you are starting to fall backward. This unbalancing may be followed by chills and goose bumps across your upper back and shoulders. Your fingers may become engorged with blood. Ideally, if this is your experience, you have altered your ordinary reality. This is when non-ordinary things can happen.

Satisfaction is not guaranteed with this offer. There are an enormous number of variables involved in transforming one's personal feeling about the nature of reality. And it *is* a feeling. Don Juan insisted that there is no reality beyond what we feel, and what we choose to feel depends upon our sensitivity and capacity for seeing beyond the obvious stage sets and hackneyed scenarios provided for us by consensus reality. Keep telling yourself, "It *can't* be real. It *can't* be real!" Yet keep in mind that conceptual understanding does not guarantee extra-conceptual change. One must engage the feeling heart to move beyond the conceptual.

Also, remember that performing the double-slit awareness experiment is not a tool, a method, or a technique. Tools, methods, and techniques are special applications of known components of consensus reality used to leverage our perception from point A to point B. Both points share the same reality, just from different perspectives. A tool or method is still a consensus reality concept based upon the use of cause and effect to produce an outcome in space-time. It doesn't change anything essentially. It just gives you another way to talk about what you still are.

In other words, techniques enable us to remain unchanged in a new way. You can't disoblige reality that way. You wind up still feeding its morphic field. Even outright opposition to consensus reality feeds its morphic field with attention, emotion, and value judgment because a morphic field does not discriminate between "good" input and "bad" input. Cultural and psychological morphic fields are omnivorous. By acting dualistically in opposing consensus reality, you validate a conflict between polarized potentials and create excessive potential for the negative polarity. This automatically feeds more potential into the positive polarity to balance the excessive potential created by your opposition. The more attention you give to one potential, the more you increase the strength of its potential opposite. Hey, it's just quantum nondualism. That's why we are learning to *disoblige* reality rather than trying to destroy it completely. It has its place, but it has to be put in it instead of letting it run wild everywhere else.

So here we are, pulsating away in a reality of our own creation and doubting its validity. There is a way of disobliging it by beating it at its own game. You become a mole within its sly operation. Let's say that you have accepted the evidence that reality is merely an illusion, developed an attitude and mood of thoroughgoing doubt about what it shows you, and then realized you can choose whether or not to oblige its simulated universe. Then you started performing your double-slit awareness experiment and finally began to make your observer redundant by acting faster than it can measure and entering a feeling state that your observer can't incorporate into its persistent process. You are ready to go behind the operational structure of the illusion. You are ready for some black ops.

5

Not Seeing the Forest *or* the Trees

Now you go undercover. Biologist Lyall Watson has written a number of interesting books using his scientific background to illuminate rather than disprove some very strange and extraordinary experiences of the paranormal in exotic and remote parts of the world. In his book, *Gifts of Unknown Things*, Watson tells of an experience with a young shamanka (a female shaman) named Tia on the island of Nus Tarian in Indonesia. At one point in his account of his stay on Nus Tarian, Watson tells of coming upon Tia and a very young girl sitting on the trunk of a fallen tree in a grove of *kenari* trees. Tia is explaining something to the little girl, but having difficulty making her point. She decides to take a different approach, and getting up, begins to dance, her movements suggesting the forms and patterns of the trees around her. Watson is made cognizant of the beauty of the trees by the expressiveness of Tia's dance. Then, by changing the form of the dance, she introduces herself into the trees' pattern. Now there are the trees and Tia among them. Watson says that what she did next is something impossible to describe:

> She seemed hardly to move. It wasn't a group of gestures or a fixed pattern of steps—nothing that could be choreographed. But nevertheless, it was real. It had little to do with dance, and yet it was the essence of all great dancing. What it achieved was to convey a feeling, to make a suggestion, and when

> it was done, we were different. The trees existed, Tia
> existed, and somehow there was a vital connection
> between them (203).

She repeated this expression of association between herself and the trees again and again with shifts in emphasis and perspective— seeing it from various angles but always repeating the same motif of connection. Through her dancing, Tia established an identity between the trees she perceived in her mind and the physical trees around her. In other words, she both internalized and externalized the trees in such a way that there was no longer a difference between the two versions. So vital was this connection that when Tia blotted out the image of the trees in her mind, the trees around her vanished as well.

"One moment Tia danced in a grove of shady *kenari*," said Watson, "the next she was standing alone in the hard, bright light of the sun." Stunned, Watson tried to regain his perception of the trees by blinking and rubbing his eyes, and slowly, the trees reappeared. The little girl Tia was instructing was delighted by the lesson shown to her. She ran around touching the trees and laughing, then stopped in front of Tia and covered her eyes with her hands, in effect blotting out the world. Then she removed her hands to reveal the world and the trees once more. The little girl understood the discontinuous fluttering of reality—on, off; on, off.

What the hell? What just happened? I only just got here. There are forces at play here that aren't supposed to be in our tidy consensus reality. Still, we can note some vaguely familiar elements that have been presented earlier in this disquisition—a pulsating reality for one, an external and an internal reality for another, a dualistic seesawing of perception from one to the other for a third, an essential link between the two for a fourth, and the creation of a feeling state for a fifth. And a new element of playfulness is added as the sixth element. According to Watson, the grove of *kenari* trees vanished, then reappeared. His accustomed reality was shaken, and I'm certain doubt and uncertainty followed. Luckily, Watson had had previous experiences of a non-ordinary nature, so he did not automatically deny what he experienced or rationalize it away. On, off; on, off. Aren't we living in a continuous reality? How can it go off and then come back on? And what is scarier, a young girl in Indonesia can make it do that, and not just for herself but for other people as well. Katy, bar the door!

Let's speculate. It's about all we can do at this point. To begin with, Tia is the product of an unfamiliar (to us) cultural morphic field. Her community is traditionally Muslim, isolated from modern secular and technological influences by its remote location, but there is a much older substratum of native tribal shamanic belief. In Watson's account, Tia upset the Muslim status quo and came to be regarded as dangerous by the locals since she was able to alter bits of their reality, specifically as a healer. But we are concentrating on her ability to make a grove of trees disappear. Tia's individual morphic field was not being shaped by the island's consensus reality morphic field in which such things could not happen. Remember this.

She was an accomplished dancer, a creative artist who was able to express certain themes and motifs as a feeling state.

She could internalize an abstraction, in this case, "treeness."

She was aware in a very subjective way of her creative relationship with an external reality. She "felt" her subjective identification with the external trees through her observer effect. She knew the trees were as much her creation as they were a seemingly independent phenomenon existing apart from her perception of them. Remember this.

Knowing the nature of her relationship to the external trees, by blanking her internal dialogue with them (she was, in effect, playing host to the trees that were in her mind) she stopped the transfer of the internal trees outward from her observer to the field of infinite possibilities and canceled their appearance as physical objects. Tia "entertained" (in the sense of "to receive as a guest") the trees in her subjective awareness but did not regard them as intrusive and compelling features of a hard, external reality. Their beauty was in their innate fluidity and infinite capacity to assume innumerable forms. Tia's trees remained tentative and incomplete as definitive, physical trees. They existed in two places at once—in her mind and growing out of the ground in front of her. The Western mind would insist on a choice between the two alternatives: either the trees existed only in her mind or they existed only outside of her as part of the physical landscape. They could not be in both places at once. Here, the quantum mechanics principle of superposition might be in play. However, the principle applies to things or events in their un-collapsed wave function state before observation has taken place, as in the case of Schrodinger's cat, which is both alive and dead prior to being observed.

Somehow, Tia was able to move her awareness back and forth between a state of indeterminate wave potential (enfolding) and its collapse into an observed reality (unfolding), and she did this, I believe, by altering her feeling state rather than by thinking about the process. Another possibility is that she alternated her thinking state with a non-thinking state, but this seems much too deliberative for Tia's instinctive and spontaneous awareness.

I think it is more likely that Tia infiltrated the pulsating frequency of reality itself. On, off; on, off. She was able to choose between the two states by using her selective attention. As Watson says, ". . . wanting something changes the thing you want. She gave up wanting and let herself change" (206). This is what I meant by "going undercover" and being a "mole" within the pulsating process of reality. Tia was not outside the process, manipulating it; rather she had *become* the process by recognizing it in herself. She felt herself switching on and off and waited for the gap between the two states, like waiting for the right moment to jump into the twirl of a jump rope being held by its ends and swung around and around by two other jumpers.

"Waiting for the gap" is hugely misleading. It is a way of translating something occurring outside of space-time into an understandable context. Tia had to interact with a process that, although fundamental to the existence of the universe, cannot be perceived under ordinary circumstances.

Friedman, in *The Hidden Domain*, cites the discarnate entity, Seth, channeled by Jane Roberts, as indicating that the flickering on and off of the universe occurs every 5.3×10^{-44} seconds, an incomprehensibly minute pulsation of time (164). (Seth, a controversial persona, has gained more and more academic credibility over the years since appearing in *The Seth Material* in 1970.) The reason Lyall Watson shared the experience of seeing the grove of trees vanish is because Tia did not cast a spell to cloud his awareness or hypnotize him into a suggestive state. She jiggled reality itself, at least as we perceive it to be, and Watson was a part of that reality, thus incorporating him into its alteration. Watson was on Tia's turf, so to speak. He was in her morphic field, not his own. This is a colossal achievement and one not likely to be duplicated within our Western cultural morphic field of consensus reality.

Nevertheless, we can install the fascinating beauty of Tia's ability in our awareness and love the possibility of its appearance in our own lives. Meaning is being. We expand our awareness to include what Tia has done and thus increase the likelihood that we will be able to do it too. We do not need to concentrate on doing it or focus our intent on making it happen. That's a sure way to keep it from taking place. Remember, we *act*. It's like the actor standing before the closed door identifying items within the room behind it. Finding the gap between the switching on and off of reality is already a probability. Given the fact that a young girl in Indonesia has already done it makes it even more probable. Just allow for the probability to become a part of your reality. Don't give it a time-frame, a historical context. It is already here now. It is the nature of both yourself and the simulated reality that appears to surround you. Notice it as part of your daily life, your ongoing awareness, and *act* on it. Remember this.

Now let's get to the fun part. Tia and her little friend were playing. Although Tia was trying to explain something to her companion, she was not engaged in a serious lecture about whatever it was she wanted to convey. She was just trying to make a point about something, and seeing that the little girl wasn't getting it, Tia decided to dance her point to make it easier to grasp. Obviously Tia was not using left-brain verbalization as her teaching mode. That wasn't working. She decided to engage her right-brain creative side in the form of dance to express a suggestion of her meaning. She wanted the little girl to feel what she was feeling rather than to conceptualize it. This proved to be an effective approach.

Watson gives us an operative context for what Tia was doing. Children are more susceptible to paranormal phenomena than adults because their minds have not yet been set into consensus reality's prescribed formula. For them, anything is still possible. Unrestricted in their perceptions and in their thinking, children draw conclusions about reality that to adults seem nonsensical and deluded. The adult view tends to prevail unless you happen to be in a playful state of mind. The adult, mired in a prescribed mode of thinking, separates his or her "real" perception of the relations between things from the child's seemingly confused and irrational perception of them. It's called "magical thinking" by the adult thought police who feel obliged to enforce the rules of consensus reality.

Watson goes on to describe the activities of a group in Toronto who wanted to experience a collective hallucination. They decided to try to create a completely fictitious seventeenth-century nobleman named Philip. Nothing happened for two years until "a session when several members of the group were involved in some childlike horse-play" (214). Suddenly, the table at which they were seated began to rock, apparently of its own volition. This gave the group a useful insight. When you adopt childlike behavior, singing silly songs and adopting a magical view of reality, you can produce psychokinetic phenomena whenever you like. Katy, bar that door again! Once more, we are given a choice of realities. In this case, all you have to do is become as little children to make impossible things happen. The catch is that you can't have all the careful scrutiny and assessment of adult "common sense" and experience the impossible at the same time. It's a trade-off. You can be the serious adult and the silly child by turns, but you can't combine the two experiences into one. They are mutually exclusive.

So, take all that you have read so far about the illusion of reality, the absence of any "real" space and time, the creation of real-looking holograms out of infinite potential, the absence of anything resembling cause and effect, the flickering on and off of physical reality, and think of them as toys to play with. If you get silly and creative with the universe, it will respond in like fashion. Let's call it "the Tia Effect."

You can't set about applying the things I've told you as a methodology for becoming a master or mistress of reality. You will remain on the outside looking in. There is a surrender required here, a surrender of all rational purpose, expectation, and anticipation of a specific outcome. Think of the amount of trust required to bring about the Tia Effect. If you want to play in her reality, you have to be willing to join in the fun purely for the fun of it. Drop the "Yeah, but ..." and "What if ..." precautions. You can do it. Just open both slits of your double-slit awareness, forget about who you're supposed to be and what you're supposed to be doing, and *act*. Have fun, and remember that *everything* is just made up.

6

Moving Your Mind-Set

The big stumbling block all along has been your programmed mind-set. Don Juan has a lot to say about changing your mind-set. He calls it your *assemblage point.* So influential has been Don Juan's elucidation of the assemblage point that a whole industry has grown up around the concept. Look up *assemblage point* online, and you'll see what I mean. You can now become a certified assemblage point practitioner. This means a certified professional can move your assemblage point from an unproductive or even destructive position to a beneficial one.

The assemblage point is your established perception of reality. Most people's assemblage points are oriented toward the perception of the *tonal,* so what most people experience is consensus reality. However, a good slap on the back is supposed to move the assemblage point to a position whereby one can perceive the *nagual* and therefore experience non-ordinary worlds and parallel universes. We are cautioned, however, that the apparent lateral movement of the assemblage point is actually a change in the depth of the point within the luminous cocoon (an egg-shaped shell of energy) surrounding the human body. We are already misled about where the assemblage point goes when it is moved. That's just the beginning. There are all kinds of provisos and an actual schematic showing what the assemblage point is affecting.

For one thing, the blow to the back, which moves the assemblage point, has to be delivered by a *nagual* or sorcerer who already has the ability to *see*—that is, one who is able to perceive non-ordinary reality. Also, the one who receives the blow must already be aware that the world he or she perceives is the result of where the assemblage point is located on the cocoon. Further, the assemblage point acts like a magnet that attracts emanations (fields of energy) and bundles them together wherever it moves within the human band of emanations. There is a risk that one's assemblage point might be moved to a region of the band diametrically opposed to its normal position, and then one is exposed to man's dark side, an unknown region that is somber and foreboding.

The situation is further complicated by the fact that one's assemblage point is located in an arbitrary position chosen by one's ancestors. It's not in the same location for everybody. Then there is a crucial threshold beyond which the world vanishes if the assemblage point is moved across it. In addition, one must know how to cluster the clusters of one's emanations by skimming the best ones and obliterating the alignment of the others. Finally, one must learn to integrate the countless positions to which one's assemblage point has moved in order to create a coherent whole. The danger from not accomplishing all this fine-tuning is that one might be eaten by the Eagle from which all the emanations originate. How inviting is this little transformational project?

You see how one perceptual framework (your mind-set) gets replaced by another that is just as complicated, taxing, and potentially dangerous. The only difference is in the context in which each of the mind-sets operate.

Let's take out the fine print and hidden requirements. Play like you have an assemblage point. I don't mean *imagine* that you have an assemblage point. Imagining isn't necessarily playing. You can imagine a lot of depressing and scary stuff. *Play like you do.* Now play like you can move it. Get into your woozy, abstracted feeling state and *feel* your assemblage point shift. Play like it shifted. If it makes you laugh, then the shift was successful.

To make the shift more likely, have a friend do it. Have him or her stand behind you and join you in your abstracted state. Remember, you are *playing* like this is happening. Have the friend place one of his or her hands on your shoulder. If both of you begin to giggle, then you are creating the proper feeling state for the shift to take place. Now have your friend say, "I shift your assemblage point *now!*" and with a finger or fingers bunched into a bird's beak, poke some point on your upper back between or below your shoulder blades, but no lower than the upper two thirds of the back.

The essential requirement for shifting the assemblage point is that it be performed as *acting.* Your friend must not hesitate or deliberate about where your assemblage point should be shifted. He or she should just immediately *know* where the point is. You should feel that drawing, pulling sensation backward and a flowing sensation down your spine and into your legs and feet. Do not try to direct what you feel is happening. Go with it. Let it affect you in whatever way it wants. Not that hard, is it? See if your reality feels different to you in any way. Don't go back to your consensus reality for an analytical survey to do this. Just find the feeling. Be with it. Let it reorient your awareness. If you feel stupid and silly (not about what you are doing but in a purely joyful, childlike fashion—as though you had spun around on a swing just to see how you felt afterward), you have altered your mind-set.

Whether this is a permanent or temporary shift of the assemblage point is not important as long as a definite shift is felt. You can repeat the process as many times as you feel the inclination. You will probably need to practice making the shift a number of times anyway in order to become familiar with it.

Remember that you are disobliging reality every time you shift your assemblage point. In effect, you are asserting your personal choice of a reality different from the one you have been programmed to accept as the only existing one. You may not be an apprentice Yaqui sorcerer, and you may not be adhering to all the requirements for Don Juan's unique shift of the assemblage point, but if you limit your practice to his particular technique, you are enforcing a single-slit reality in preference to a double-slit alternative.

Remember the Tia Effect. As long as you are playful in your approach and generate suggestions and convey feelings rather than apply a prescribed method, and as long as you allow processes to flow and to take their own courses, you are in the gap between on and off "where forces are beyond comprehension."

This whole process of *disobliging* reality involves a complementary process of *detaching* from reality as well. Don't think so much of being proactive or taking action to fix, change, or improve your situation. You are cutting ties, letting go of entanglements, saying your good-byes to old acquaintances, and preparing to move on. You are unencumbering yourself of pretty much everything that holds you to a consensual space-time identity. How you feel about your many-faceted experience in reality—the worries, the incentives to achieve this or avoid that, the expectations for happiness and success, the troubling relationships with others, the desire to please or be accepted and validated by people you admire and love, the hurts and disappointments you have suffered, the addition of more and more apps to your smartphone—these are written in your face, your expressions, the look in your eyes. There is a purer you. Less is more. You are learning to travel light into the light. Remember this.

7

If It Walks Like a Paradox and
Talks Like a Paradox ...

Now, if at this point you think what you have read so far is more than enough to chew on, stop. Don't continue. You will only get increasingly addled. We are headed into murky waters with an unreliable compass. Remember the uncertainty principle. That was about location and velocity. Here we are trying to find something similar but within a much broader context. While you may determine what map you are using, you can't be sure if the map is accurate, and if you know that the map is accurate, you can't be sure that what it is showing you can be determined. Hang on. It's going to be a bumpy ride.

Essentially, we will be entering the Bizarro world. Remember the Bizarro world from the Superman comics? In that world everything is backward. Wet is dry, up is down, short is long, far is near, and so forth. We think that by using our standard of reality, we can determine when things are backward and therefore part of the Bizarro world. We can formalize our reality and recognize what is opposite to it and therefore identify what doesn't fit. It's all based on dualistic thinking in which things exist as opposed pairs or as polarized opposites. That's our comfortable standard for reality. How can you know for certain what is up unless you have a clear idea of what is down? How do you know what is dry unless you have felt what is wet? It is a very divisive kind of reality that separates the sheep from the goats: sheep over here; goats over there. When the difference is not so clear, we get separation anxiety. "If I can't tell where the sheep are, I'll never be able to find the goats. I might not even be able to tell which are which, and then I'll be stupid and lack common sense and be a disgrace to the only reality there is." Pity.

Learning is supposed to be a highly valuable commodity, both as a noun and as a verb. Even in the afterlife, according to those who have crossed over briefly and returned, having committed a significant part of your physical life to learning is considered laudable. You have put your gift of physical existence to good use if you have improved and enlarged what you know. I'm fairly sure, of course, that a lot depends on the kind of learning in which you engaged. Having mastered the art of flimflam probably won't earn you any stars. I believe self-development is the preferred area of instruction. Spiritual improvement seems to be the subject area most prized. I'm not being flippant about this. I regard it to be true in the sense that its level of meaning is among the most comprehensive.

Now think about your most exciting learning experiences. I'm not talking about completing degree requirements or being praised or rewarded for understanding a recognized area of expertise. I'm talking about the thrill that went through you when you realized you had grasped new knowledge, learned something you hadn't known before that had a major impact on who you are. This would be a life-changing kind of learning experience. I am referring to suddenly being made aware that what you formerly believed to be true is actually the complete opposite of the authenticated certainty. That's the biggest thrill of all.

Most people's idea of learning is the accumulation of more and more information about a subject or skill. Accretion, however, isn't particularly thrilling. Finding out that your conviction about the truth of something is worthless because it is wrong, and not just worthless but worthless squared because it asserts the opposite of what happens to be the actual reality, gives learning a huge surprise factor plus the thrill ride of having to reorganize what you thought to be True with a capital T. It can be a transformative experience. Let's go there, shall we?

In 2009, a DVD entitled *The Living Matrix* was released. It is about "the new science of healing." Various philosophical and alternative healing authorities appear, along with subjects who report their personal experiences of non-ordinary healing. One of the subjects is named Arielle Essex. She was diagnosed with a brain tumor after having consulted a doctor about her recurring, severe headaches. The doctor discovered that Arielle's hormone balance was seriously upset by the tumor and that one of the effects of the hormone imbalance was to make Arielle incapable of having children. Ironically, her greatest wish at the time was to have children. She had recently divorced her husband because he was opposed to the idea. Now she was faced with taking steps to remove the tumor. She refused conventional medical treatment and chose to use neuro-linguistic programming to treat her condition as she was already a master of the practice.

Note that Arielle was conflicted about her ability to have children. She desperately wanted them but now was rendered incapable of conceiving them. She embarked on a rigorous program of research into all aspects of her condition, gaining knowledge from a variety of traditional sources and from personal interaction with people who had some acquaintance with her condition, either as medical professionals or as fellow sufferers of its effects. This went on for ten years.

Arielle was desperate. She adopted the unwavering attitude of "I want to be rid of this tumor!" Yet her condition remained the same. Weary of fighting, Arielle began to entertain the idea of accepting her situation. Significantly, she phrased her surrender in the form of an open-ended question: "What would it be like if I accepted this tumor?" The importance of the open-ended question is that it does not imply a specific answer. There is no consideration of the most likely treatments and cures. The question asks, "What would it be like?" This leaves every possibility open, from crossing your eyes to moving to another country.

Arielle began to reflect on her tumor and its implications. Her experience hadn't been all bad. She had learned an awful lot, met many interesting people, and certainly got to know herself more thoroughly.

She began to wonder if her tumor had a purpose as yet unknown to her. She finally gave it permission to remain with her for the rest of her life. Basically, *she gave up*. Remember this. In six months, it had disappeared. In the process of actively fighting and then dealing with her tumor in various ways, she had become a different person. Her doctor mentioned noticing that specific change in her. She was no longer tumor-inducing Arielle; she was tumor-free Arielle because she was no longer the Arielle she had been.

Bear with me a little longer. I have another example similar to Arielle's but much more mundane. This will give you a better idea of the scope of the point I am trying to make.

Jeremy Wade is a world-class fisherman who stars in a TV series called *River Monsters*. He travels the world looking for big, dangerous aquatic creatures he hopes to catch in order to solve mysteries in out-of-the-way places about water monsters that kill people. In a recent episode, he was in Canada trying to catch a muskellunge, a huge pike that may have been responsible for creating a pervasive legend about a serpentine monster that from time to time attacks humans. Now, if anyone knows how to catch fish, it's Jeremy Wade. Jeremy selected his heavy tackle and a lure that actually made a noise as it was pulled through the water. He figured that the muskie would be particularly drawn to a noisy lure. He began to fish. The waters he was fishing in were crystal clear, so he could see the muskie as it tracked his lure. Over and over again, the muskie followed the lure to within inches but never bit. Jeremy was nonplussed. He fished from dawn into the night day after day without a nibble. I believe he kept this up for three days. Finally admitting defeat, *he gave up*. Remember this.

To console himself, he decided to go back to fishing for small fry just to reassure himself that he could catch something. He got out his light tackle and light line and cast into the water. Almost immediately, a huge muskie took his hook. Jeremy now had to try to land this monster with tackle that was never designed to withstand the weight and violent fight of such a big fish. It took him a while, but his skill and experience enabled him to bring in the muskie.

My point is obvious. Both Arielle and Jeremy got what they wanted when they gave up trying to get it. The irony is apparent and inescapable, but the paradox is practically sublime. In our reality, success is not based on quitting. We are taught never to give up, that if you want something badly enough, you need to never slacken your effort to get it. You struggle and fight until you win. Otherwise, you are living in a Bizarro world where winning is accomplished by losing and nothing makes sense anymore.

Note that both Arielle and Jeremy were in a conflicted state, Arielle because her desire for children was thwarted by the appearance of a tumor that made it impossible for her to have them and Jeremy because, although considered a master fisherman, he couldn't get the particular fish he was after even following days of unrelenting fishing. What they shared in their desperation was a very focused intent. However, their unwavering intent was followed by a complete letting go of their respective goals. What's going on here? Why is everything back-assward?

Consider that intent, when highly motivated, as in fear of death or loss of reputation, requires that you hold onto something with your consciousness and your will. You are setting up a blocked or static state in which nothing will change. When you give up, the intent you so intensely created begins to flow, allowing for whatever outcome is suitable to the situation. You are no longer directing anything or trying to make a specific outcome happen. You have surrendered your designing awareness to an abandonment of all awareness of what it was you wanted to take place. This is quantum creative potential in action. You stop polarizing the situation and move to a neutral state where unknown forces can operate. Trust is paramount here.

Remember, on the quantum level, you are bypassing cause and effect as well as time. You have an equation: intent, letting go, and trust. Think about incorporating that equation into your daily life. Sure, it's missing some of the elements our reality considers essential to successful accomplishment, but why ignore a functional possibility just to support the authority of a habitual, and therefore limited, principle? Despite the fact that scientists do it all the time, we have the freedom to play with what's available, and that includes limitless possibilities.

The reason things work in a Bizarro way is because of the risk of creating excessive potential and inflating morphic fields. Consciousness rocks. If you aim it in certain directions, it not only creates physical reality, but it also causes that reality to become increasingly pervasive. Its persistence increases exponentially. Recall my cautionary advice about using your double-slit awareness experiment as a tool or method. If you are using space-time components to reduce or eliminate what space-time is doing, you get nowhere. You create excessive potential and tip the balance of polarities against yourself. Too much attention given to a polarity for too long energizes its opposite polarity.

The Taoists were singularly wise to create the *Taijitu* or graphic illustration of the interaction of polarities. Inherent in the action of the yang polarity is the latent action of the yin polarity and vice versa. They are always interchanging. Balance is the ultimate integration of the two forces. When Arielle and Jeremy gave up their struggles and let go, the opportunity for balance was restored. It's all in knowing how to finesse the forces of the universe.

Let's take a very popular activity from our consensus reality as an example of building excessive potential. Campaigns against widespread serious diseases or destructive behavior always manage to rally masses of people to their cause. Marches, runs, or symbolic pilgrimages are held to show opposition to whatever the current health crisis or behavioral excess happens to be. Colored ribbons are devised and displayed prominently to declare supporters' commitment to eliminating the problem of choice. Fundraisers are organized. Political action groups are formed. Every attempt is made to raise public awareness of the selected crisis. Victims are made heroes, and enabling treatment facilities are lavishly endorsed. Considering what you have learned so far about quantum wave potential and morphic fields, this type of activity is just plain scary.

First, cause and effect are invoked as the central dynamic of whatever campaign is in play. Then a polarity (always a "good" one) is selected and fed enormous amounts of attention, emotion, and effort, thus creating an ever-growing morphic field of mass participation. Opposition to this field is virtually impossible as it is, in itself, defining an aspect of consensus reality. Notice that the campaign to "find a cure" is always after the fact of the disease or the behavioral aberration. It is always reactive. The strategy is: "Well, now we've got this thing to deal with; how do we fight it?" Typically, much study is given to the specifics of what the thing is, its history, and what it is to be called. These elements anchor the thing securely in our reality, along with our fear of it and the terrible things it can do. Anyone suggesting that the disease or social problem might just be a metaphor for a non-physical imbalance of subtle forces would be run out of town on a rail (a practice that retains some romantic appeal for many these days).

Aside from the money to be made by pharmaceutical companies, laboratories, clinics, and treatment centers in combating the scourge, the "fight" against it, by loading excessive potential into its opposite polarity (an increase in the spread of the condition or its development of a strong resistance to its treatment), ensures that the condition will not be eradicated anytime soon. My point, of course, is that the mounting of enormous effort and resources to destroy a polarity only strengthens the quantum potential of its opposite polarity and may well bring about the victory of the "enemy" being attacked. Our consensus reality is easily duped by the unacceptable as well as unbelievable behavior of the Bizarro world.

This course of inquiry could develop into a tangential exploration that might go on indefinitely. I see no need to go there, and I imagine it would become trying to you as well very soon. Let me just dispense with the fairly predictable objection to what I have just explained. Skeptics call on examples of the successes of conventional medical practice to show that nonsensical notions of noble endeavors creating their own failure just aren't so. Dr. Jonas Salk's dramatic cure of polio is a favorite bludgeon used to silence the more metaphysically gullible (we doubters). The following information comes from Lynne McTaggart's book *What Doctors Don't Tell You* (114-115).

In 1953, Salk announced his development of a vaccine that would prevent polio. Mass inoculations followed. Nonetheless, cases of polio increased by 50 percent between 1957 and 1958, and by 80 percent from 1958 to 1959. Then the medical industry cooked the books in reporting how many cases were actually appearing by calling polio by another name—"viral or aseptic meningitis" or "cocksackie virus." Thus, *polio* diminished while it continued to spread under other names.

The reason cases of polio are comparatively rare today is because the disease, which is cyclical, ran its course. Salk just happened to develop his vaccine toward the end of the cycle, which had begun in the late 1940s and early 1950s. It was a noble endeavor that happened to coincide with a natural decline in the virulence of polio. In this case, creating excessive potential may have been synchronistic with a result that was already in progress prior to Salk's development of his polio vaccine. So the campaign against polio didn't develop into its opposite (although cases of polio increased while mass inoculations were taking place), but the campaign proved to be incidental to the eventual outcome. Had polio not begun to decrease on its own and the campaign against it had intensified, the Bizarro result might have been a dramatic reversal of the desired outcome—or not. Remember the uncertainty principle.

If you are surprised and your convictions have been contradicted by the sorts of experiences described above, you are being drawn into more personal contact with polarized opposites and paradox. From the start, I explained to you how you are creating a reality that isn't real; how time and space are illusions; how consensus reality is actually discontinuous and oscillates interactively with another reality; and how other worlds and alternate universes exist that are just as real as the one you believe you are living in even though you can't perceive them. These contradictions and paradoxes have a limited impact on your notion of reality because, in its persistence, you continue to experience the consensus reality you are used to. Personal accounts of non-ordinary Bizarro experiences are more convincing. People who have been there/done that introduce their own evidence of Bizarro experiences, making them more than theoretical. Hopefully, once you begin to disoblige your reality, you will be more open to accepting what sounds purely theoretical as genuine, reliable information that affects your personal experience.

Here's a final contradictory paradox: chaos theory. Chaos means "utter confusion." How can you formulate a theory of utter confusion? Well, let's begin by once more noting that your dualistic interpretation of "chaos" is just another label, nothing more. In the nineteenth century linear theory prevailed and worked pretty well when applied to linear systems. However, once it was acknowledged that there existed systems that were non-linear, linear theory had nowhere else to go. Then came quantum mechanics, which opened the door to non-linear reality. Before quantum mechanics, people believed that things were caused by other things and that all dualistic oppositions were sound. Thus, by identifying every component of the universe, we could predict what would happen no matter how far into the future it occurred. Mystery and uncertainty had been explained.

Much of what continues to be accepted as reality is still based on that supposition. Wrong squared again. Not lines of force but patterns are what dominate our reality, and those patterns are stimulated by the sum of many tiny pulses. Sound familiar? Ever hear of the Butterfly Effect? That comes out of chaos theory and says that the tiniest variable can have an immense effect on wider circumstances precisely because the pattern created is nonlinear. The lesson here is that we really don't know what we're doing when we do anything at all.

8

Un-leveling the Playing Field

Now is a good time to return to Norman Friedman's explication of the function of levels of consciousness in *Bridging Science and Spirit*. He says consciousness is about focus. Citing Ken Wilber, Friedman calls the lowest level of consciousness "insentient consciousness" and says that there are many levels above that that are "less dense and more wholistic." The levels, he adds, "are not separated, not discrete, but mutually interpenetrating and interconnected." It is focus that divides these levels in consciousness. To move up to a higher level, we just need to broaden our focus. "Then," Friedman says, "the higher world becomes manifest, and the consciousness exists on a new plane" (Friedman, 99). Hey, isn't that what disobliging reality is all about?

But here's the sweet stuff. Friedman continues, "While a higher level contains the attributes of the lower level, it also has new aspects clearly different from those of the lower and cannot be seen as a derivation of the lower plane. This notion is in contradiction to the reductionism of nineteenth-century physics" (100).

The point that tickles me so much here is that consciousness does not function in the same way as learning and doesn't follow the law of accretion. It's not about adding to or increasing a store of the known, but changing the focus of your consciousness so as to make it more inclusive as awareness itself.

Citing Wilber again, Friedman says that what happens is that at each level of consciousness "an appropriate symbolic form emerges and mediates or assists the emergence (through differentiation) of the next higher level" (102). In effect, there is an evolution of consciousness to the next higher level, which transcends the lower one without obliterating it so it can use the higher structure to continue to operate on the lower but obviously in a different way. The consciousness no longer *identifies* with the lower structure and *disidentifies* or detaches from it. Remember *detaching* from the reality you were taught is the only valid one? The higher level can then integrate the lower one and appreciate (figuratively speaking) what a silly numbskull it has been.

But the really sublime aspect of this contradictory paradox is that, when viewed from a lower level of consciousness, the higher one appears to be *chaos*. It's not until our consciousness moves to the next highest level that it can perceive its order and beauty. Before then, nothing computes. The higher level makes no sense. We are strangers in a strange land. And this hierarchy of levels of consciousness is infinite. We never get to a definitive level of consciousness. "Why bother?" I hear someone ask. Boredom. It's built into our consciousness from the start. Curiosity is as well. The problem for slack asses is that consciousness eventually wants to experience more of itself. There's some kind of thrill there, but it's nonphysical. Instead of the addiction to the adrenaline rush, we develop an addiction to the bigger-focus rush. Everyone proceeds at his or her own pace, but the motivation never disappears entirely.

If the next-highest level of consciousness appears to us on the lower level to be chaos—such as giving up to get what we want—and nothing makes sense anymore, we avoid it simply because we can't get our heads around it. We fear a major dislocation of ourselves from what we have come to regard as "essential reality." We can't appreciate the idea that our reality is made up of attributes *contained* within the seemingly chaotic level above us. It's just a matter of refocusing and learning to be at home with new customs and new language. It's traveling in the very best sense of the word, and traveling, we are told, is broadening.

Let's assume you have given up and decided to move to the next-highest level of consciousness where giving up gets you stuff.

Using Arielle and Jeremy as examples, you assume that you have to be in an emotionally conflicted state to initiate change. This would imply creating a lot of urgency and discomfort for yourself in order to kick start your placement of intent. That's an unwieldy train of cause and effect to manage.

Recall that on the quantum level, space and time are illusions. You don't want to make a measurement (collapse wave function) in the usual make-a-world way that you have been used to. You don't want to build excessive potential and risk getting the opposite of what you want. Instead of forcing the issue by becoming self-conflicted or engaging in an intense struggle to change circumstances, you *allow* for something to be different. That's your intent. You can tag it with a general orientation, such as healing, resolution of conflict, improved finances, or a convenient parking place, but you don't want to get too specific. Fuzzy, isn't it? Well, you're dealing with infinite possibilities, so how are you supposed to pick one without making it collapse into the usual hologram you're already surrounded by? If you get too specific, you're back at the sorting tray, and your options are limited.

Say you want a new car. The manifesting gurus recommend that you stick pictures of the car you want onto your refrigerator and wander around picturing it in your head all the time. This is metaphysical cheerleading. Lots of spirit, but you won't necessarily win the game. No. Instead, create the feeling, the experience of being in the new car (unspecified), driving around, and getting a lot of joy out of what you're doing. Every time you do this, get into your abstracted state so that you enter a separate reality. It's enormously different from being in your left-brain reality, sizing everything up, and calculating how it's supposed to work. So you're in your car (unspecified) and in your dopey state (it's okay to be driving that way since you're in a separate reality from this one), and you feel the flow of abstracted awareness down your back and legs and into your heels. Cut it off. You've had your fun. Forget the whole thing, and I mean *forget* it. Don't give it another thought or sensation. Find something else to do completely unrelated to manifesting the car.

Is this another Bizarro contradiction? It would seem so. But you see, you need to interrupt continuity. Continuity is space-time reality. Continuity is cause and effect. Continuity is polarization and the creation of excessive potential. What you want to happen is that you catch the gap. Remember the gap? The gap is the pause between on and off. When you maintain continuity, there is no gap because you perceive everything to be on continually and then you don't allow room for something to be different. All you will get is what being *on* allows you to have, which is only what you already know about in the on state.

What you want is coherence—coherence with infinite possibility—and that is discontinuous. *You* are discontinuous. As a result, you have to find the gap between who you assume yourself to be (a continuous conscious identity) and what you want (which isn't continuous either). That's why you place intent in your *on* mode and let go of what you just did in your *off* mode. But you flip the switch so fast that space-time doesn't have sufficient opportunity to get into the process and gum everything up.

Be coherent with the discontinuity of the universe—always. There's no point in agonizing over whether you got your driving-the-car feeling right. That's left-brain assessment. There's no point in worrying about whether you were open to all possibilities. How would you know? You couldn't imagine what they are. There's no point in continually checking yourself to see if you have truly let go of your catch-the-gap exercise about the car because then you haven't really let go. You see, it's very much akin to *acting*. When you *act*, there is only the act. Nothing more. Meaning is being. Be the meaning of what you are. You are on, and then you are off, and then you are on again. How many gaps do you need before you catch one?

9

Love Is a Frequency, So Resonate

I've deliberately postponed introducing the subject of love. It means too many things to too many people. I thought I might cut down the size of the subject by keeping love in a quantum context as a field phenomenon. In that context, the field of all fields is supposed to be the zero-point field, which, since it is the origin of everything in the universe, would have to contain love. Except that I am enough of a dissenter to propose that love—the universal love—contains the zero-point field. No, I'm not going to get all hippy-dippy, touchy-feely, huggy-kissy, and lovey-dovey. That's just some of the various ways sentient beings interpret love with the big L. Love is universally associated with the heart as the seat of its emotion and expression. Thankfully, there is a physics of the heart, which can keep the gooey part from getting out of control.

The following information is easily referenced, so I'll just outline some of the major points. The electromagnetic field of the heart is a torsion field in the shape of a torus. The torus resembles a spinning doughnut on its side but with a zero-dimensional point called a vertex at its center instead of a hole. The surface of the torus folds in upon itself, and all points on that surface flow together into the vertex. Any input into the vertex is folded and rotated inward and then spread out over the surface of the torus. The field it creates is approximately five thousand times more powerful than that of the brain. The field extends outward from the human body to a distance of twelve to fifteen feet.

Every beat of the heart sends information to all parts of the body, which regulates the proper functioning of all the body's systems. The heart maintains a vital connection with the right hemisphere of the brain in an exchange of both information and feeling. Thus, heart-centered awareness draws on a complementary interaction between the heart and the right brain. The left brain is usually preoccupied with sorting, classifying, and retrieving information. It can be made more sensitive to the right brain's activity through the feeling heart, but most people pay little attention to what their hearts are telling them so long as the left brain is drowning out the heart's subtler hints and suggestions with "real-life" blather. So where's the love?

Well, the heart doesn't create it, but it processes it out of the universal consciousness in the form of a scalar field. Scalar fields have magnitude but no vector or direction, which means they simply fill up whatever space is available. They aren't directed or beamed anywhere, and unlike electromagnetic energy, they don't lose energy proportionate to the distance they might have to travel.

Love doesn't go anywhere because it isn't anywhere it hasn't already been. It's everywhere at once in equal measure. Thus, you can't be "lookin' for love in all the wrong places" because there is no place where love isn't. I could compare the frequency of love to "the force" in *Star Wars* except that "force" implies something being exerted in the form of work, which is the definition of energy, and love, believe it or not, is not an energy. It is a non-local, non-spatial, structured field frequency. Love is as pervasive as the force, and as accessible, but it is not a tool or a weapon. You can *fall* into love, but you can't *use* it. It's bigger than that.

All of the expressions of love used by humans are their individually limited versions of what they interpret love to be. There are no edges to love, no handles, no boundaries. Love is more than any individual consciousness can apprehend. It can be felt to various degrees according to an individual's capacity, but there is no way to get outside of love so as get a sense of its size or shape. It's more like the ether or the "quintessence" of alchemy. It is universally pervasive. It is personal and at the same time infinite. So, in relation to everything

I have told you about so far, it is *the* major player. Everything is one; everything is intermingled. Nothing is or can be without drawing on the frequency of love. You will find it in your heart—if you go there. This is both easy and terribly hard—another paradox.

Your heart is forever speaking to you and guiding you, yet most of us are tone deaf. As a frequency, love has a tone, but you have to develop the sensitivity to recognize it. Left-brain yadda-yadda drowns out the frequency of the heart. By left-brain yadda-yadda, I don't mean the usual ego-driven "internal dialogue" so maligned by teachers of meditation. That's just surface noise. I'm referring to the ideational framework that convinces you meditation would be good for you in the first place. Right away, you're into dualistic thinking, separating the sheep from the goats, feeling imperfect, and looking for a method to leverage you into becoming someone who can recognize the frequency of the heart.

You see, once you accept the notion that you need to hear the frequency of your heart, you set about *trying* to do something about it. What would Yoda tell you? "Do or not do; there is no try." Isn't this *acting*? You bet your booties. I know this sounds glib and snarky, but you just need to stop *not* listening to your heart. And where can you do that? Why, in a separate reality, of course. So now you have even more reason to disoblige the localized reality that is giving you nothing but yadda-yadda.

The heart *is* a separate reality. So is your right brain. Look at Jill Bolte Taylor's book again. You already know how to detach from consensus reality, run your double-slit awareness experiment, shift your assemblage point, and bypass your observer. These are all very important preconditions for finding the frequency of your heart. Sometimes, a *chance* meeting (*yeah, right*) with an unusually perceptive person, or a shock, or a realization that bumps you up to the next highest level of consciousness will surprise you with the discovery of your heart's frequency. Most often, we are continuing to look at the rear end of the horse in front of us.

Stand up and get into your abstracted awareness once more. Let yourself relax. Now visualize, which really means *play like*, which in turn means *be as though* (you are *not* doing the imagining technique) there is a brilliantly glowing ring floating a short distance above your head. You already know what color it needs to be. The ring is large enough to encircle your body, but not as big as a hula hoop. It is spinning. You know immediately whether it is spinning clockwise or counterclockwise because the ring is actually your awareness. You can feel the ring's glowing intensity. Focus on it for a short while. Then you absentmindedly reach up with either hand and lightly tap the top center point of your skull with your middle finger. At the same instant, drop the ring, meaning let go of your awareness.

You can feel the ring fall past your eyes downward over your body and past your feet, where it spreads and expands outward toward infinity. You feel as though you are dropping downward along with the ring, as though the ring has drawn a wave of feeling downward over your body. You have let go of the ring completely. You have let go of your ordinary awareness completely. Because you have opened your heart and become totally vulnerable, you may feel a strong emotional reaction. You may laugh or cry. You may fall down and feel suddenly more expansive than you have ever felt before. It's all good.

You have experienced the frequency of your heart, sensed its tone. You have engaged the frequency of love. If the experience has been unmistakable for you, you will probably want to stay with it for a while. Everything will feel soft and lovely. If there are people around you, you may perceive a glowing radiance about them. Whatever their physical looks may be, you will see the inner radiance of their spirits.

Perhaps the most important aspect of this opening of the heart will be your unexpected love for yourself. You will sense your own inner radiance, your own precious presence. Without that inner recognition, you cannot truly love others. When you love yourself, others can feel it, and when they can feel it, they share the love. It's not a matter of rushing out into the world and loving everyone, serving them, and being kind and compassionate toward them. That's *doing*, not *acting*. You are *implementing* love when you make that choice.

You are making a tool of it. Keep your own loving integrity, and let it play out. That's what helps others the most. That way you don't have to figure out who needs what and how much of it. Be who you are independently of others' needs and wants. Love yourself as honestly as you would if there were no other people in your reality—because, in a cockeyed sort of way, there aren't.

10

Is This as Real as You Get?

Don't be frightened. I'm not going off into a tedious discussion of solipsism (belief that your own awareness is the sole object of its knowledge). If there is non-localized consciousness, then there is something outside of or more than your localized consciousness. It just isn't what you make it out to be. The atomic "wedgie" here is the existence of other people. Who the hell are they, where do they come from, and why are they here? It would be nice to invoke the uncertainty principle and be done with it, but the uncertainty principle was formulated by another person who seems to have existed outside ourselves, namely Werner Heisenberg, thus making the uncertainty principle itself uncertain.

We interact with other people. Are they complex illusions in holographic form that we collapse out of quantum wave function, or do they have an autonomous existence independent of our own? The latter case would seem more likely because other people are manifestly different and separate from ourselves. Yet from the non-dualist point of view, we learn that everything is one and that the perception of separation is an illusion. We are them, and they are us.

Where do we go from here? It occurs to me that we can't resolve this question of the reality of other people by assuming that our consciousness is the sole arbiter of what can and can't exist. It's bigger than that. We encounter other people but as self-referencing mysteries that capture our attention. It seems true that other people appear to us as real to varying degrees. Some seem more real than others, and the least real are dismissed as "phonies." Of course, phoniness refers to another person's adoption of a pretense that we consider unfounded. Still, if you prick a phony, he or she will inevitably bleed.

But if consensus reality is essentially just decoration, what are other people? Most of the time, they are part of the decoration. They are extras in the movie or stage play of our lives. We attribute feelings and motives to them without ever checking to see if they have them. We assume that they do, that they are just like us. This is a whopping presumption. There are too many other people in the world to learn much of anything specific about them. They are like stand-up cardboard cutouts placed all around singly, or in small or massive groups to give an impression of general presence, but if they are the particular focus of our attention, they morph into specific persons with whom we can interact in many ways.

It is fairly typical of a medium that, when he or she is communicating with someone "on the other side," the departed one is said to have "stepped forward" to make the communication. Curious, isn't it? Why did the one who has passed need to step forward? Forward from where? Where was this soul before it felt compelled to step forward? Apparently, communication in any direct way is unlikely unless the departed takes that step forward.

That's my suspicion about other people whom we believe to be alive. They are part of the background decoration until they step forward. They are out of focus, contributing to the prevailing atmosphere of the place, wherever it may be, until they step forward. This is in line (or in orbit) with a somewhat marginally spiritual idea that other people should not be helped unless they ask for it. This caution helps to prevent meddling in someone else's quantum reality. Remember the Butterfly Effect? Helping other people unasked involves intruding into the reality they are creating for themselves as part of their own personal scenarios, which we gratuitously assume actually exist. The effects of such nosiness are incalculable.

Again, my suspicion is that other people aren't actually real until they step forward. We predicate them on the basis of our own experience of consciousness. Until they enter our sphere of consciousness by altering it directly in any number of possible ways, they remain out-of-focus decoration. I am talking about personal interaction with specific members of what is otherwise a collection of extras being used to enlarge a parade. Others are as real as we take them to be. It all depends upon the three Rs: reality, relevance, and resonance. If we assume they are real, then they are capable of being relevant to our experience, whereupon they may actually resonate with us in a very significant way. So we are back to our own interpretation of reality.

Robert Scheinfeld, author of *Busting Loose from the Money Game* (which isn't mainly about money at all—or at least not much) argues that other people "have absolutely no power, independent existence, or independent decision-making authority" within our individual holograms. Within our holograms, he says, "they're 100 percent your creation, and that's all you concern yourself with. You leave *their* holograms alone" (58). Notice that Scheinfeld is placing the influence of other people within the context of a holographic reality. This means other people are the result of the observer effect and the collapse of the wave function. What other people might be outside of and before the creation of a holographic reality, he doesn't say. What he does is to invoke the quantum physics concept of *tangled hierarchy*: "What it means, as I choose to interpret it, is if you try to resolve certain riddles from a logical or analytical perspective, it results in an endless loop that gets you nowhere" (57).

Think of M. C. Escher's drawing of two hands drawing each other or the answer to the riddle, "Which came first, the chicken or the egg?" The tangled hierarchy results from an inability to sort out cause and effect. What you get is a muddle of improbabilities and contradictions that beggar the imagination. Apply the tangled hierarchy to other people, and you are faced with endless loops like, "Am I creating other people who are creating me so that my existence depends upon my creation of their creation of me?" Whatever the case, another person has to "step forward" into the tangle to begin any hierarchy at all. As Rick Blaine (Humphrey Bogart) lamented in *Casablanca*, "Of all the gin joints in all the towns in all the world, she walks into mine."

But Scheinfeld offers some guidelines for dealing with other people who are part of your holographic reality. He says:

> "All the patterns in The Field [the Zero Point Field] relating to other people are created to allow them to play one or more of these three roles in your hologram:
>
> 1. To reflect something back you're thinking or feeling about yourself or a belief you have
>
> 2. To share supportive knowledge, wisdom, or insight with you
>
> 3. To set something in motion to support you on your journey" (59).

In other words, other people in your holographic reality can be useful, and it sounds like they can be useful in helping you to disoblige reality. The twist here is that you can use the other people of consensus reality to help you diminish the effect of it on your consciousness. By regarding other people as components of a larger purpose, you can eliminate the persistent tendency to involve yourself in situations that breed drama, and drama always fosters an acceptance of the prevailing reality.

So much for other people in your holographic reality. What about outside that reality? Obviously, that depends upon the degree to which you participate in a non-holographic reality. The "truth" is that the more non-holographic your reality, the less contact you will have with the people inhabiting your holographic one because that's the only reality they know. The only other people with whom you will remain in non-holographic contact will be those with whom you have created a "bridge of understanding." This bridge will establish a very strong bond between you and others who resonate with you because it recognizes the existence of two very different ways of experiencing two different realities and the choice you have between them. What you share with these nonbelieving outsiders is that "next highest level of consciousness" that still appears to be chaos to those confined to a holographic reality. The implication is that the reality you share with others of like mind and heart (they're not really separate) will forever be free of a lesser reality because it *contains* it. Therefore, you can use whatever elements of it you find useful without it having any controlling influence over you at all.

The challenge now is having to negotiate at least two realities (one of which is a portal to innumerable other ones) and continue to function normally enough not to be victimized by the prevailing reality. It's a real tightrope act. You're neither fully here nor fully there. You have to reduce the persistence of consensus reality in order to have access to a separate reality that offers unimaginable possibilities but is hampered by that very persistence.

Are we conflicted here? Very much so. The tension can be agonizing. But if you can find the *frisson* of that tension, you can keep it in a fluid state so that it does not create an outright blockage of creative potential or an occasion for generating excessive potential. *Frisson* is French for "shiver" or "thrill." Originally it meant "friction," but it usually refers to a moment of emotional excitement. What we are looking for is the thrill of disquiet or the emotional excitement of a disquieting state. We want to experience the *fun* of frisson. What that disquieting state constitutes is *intent*. It is a profound stimulus for *acting*.

Frisson does not involve forethought or implementation. It is the friction generated between two modes of being that are both exclusive and complementary at the same time. More paradox. Paradox itself can be thought of as a bridge because it contains a seeming contradiction that requires two mutually exclusive components in order to create a state of paradox, which in turn produces a frisson of disquiet. Disobliging reality is a paradoxical act and at the same time an inclusive one that predominates as a *feeling state* of being "between the worlds." It is a space, a frequency that we doubters hold and are compelled to anchor in consensus reality as an agent for shrinking the development and growth of deadening, constricted morphic fields.

Eckhart Tolle talks about "The Frequency-holders" in his book *A New Earth* (306-307), a fact of which I was unaware (at least consciously) when I began writing *Disobliging Reality.* The bridge of understanding that you share with another person is constructed of that frequency and that feeling state of being between the worlds. The bridge of understanding transcends holographic reality precisely because it is not contained by it. Rather, it contains both holographic reality and the separate reality that is outside of it.

When I say "contains," I don't mean that both realities are fully understood by those who share the bridge of understanding. To fully comprehend holographic reality, of which you are a part, you would need to be free of it, and full comprehension of the separate reality is not yet within your grasp. Still, you are aware that both realities exist in a paradoxical combination of pulsating interaction and that you are given the freedom to choose between them. As a result, your consciousness has moved to the next-highest level because of that awareness.

I know I can run, but I can't hide. I brought up the subject of whether other people are real. Certainly, as Scheinfeld has asserted, they are real as part of your holographic reality, but as you disoblige that reality and detach from it, are they still real? Are they real in a separate reality? My conclusion is that they become more real as they cross that bridge of understanding with you. Other people who are embedded in a holographic reality are as real as the other people in the same reality perceive them to be. To those between the worlds, these other people are as illusory as the holographic reality they are creating.

However, when a holographically generated person "steps forward" onto the bridge of understanding (this is the big step forward), he or she adds more dimension to his or her reality, which only intensifies as the person moves to each of the next-highest levels of consciousness. We have a baseline of nonlocalized consciousness that creates the entire universe, and this includes other people.

Yet, "In my father's house there are many mansions." I take that to mean that there are as many levels of consciousness as there are other universes to be perceived at those levels. People who report their near-death experiences insist that the wondrous reality they experienced on the other side is "more real" than this one. Why? Because it isn't limited to a three- (nowadays four-) dimensional holographic reality. It's way bigger than that. The point is that other people who are no longer constrained by consensus reality are more "real" than those who are. It's all a matter of perspective. Other people are real, but to different degrees when seen from different viewpoints. I suppose the most discerning question you can ask another is, "Is this as real as you get?"

Inevitably, the matter of historicity enters the question. Remember, everything collapsed out of quantum wave function materializes with a ready-made history. So do other people. So do we. Recognizing the reality of others depends to a great extent on knowing their histories. That's how we know who they are. We also know who we are based upon our past, which we believe to be a continuous narrative of our lives from our origin to the present. Uncomfortably for our supposed identities, our past is no more fixed than our future, and we know from quantum theory that our choices in the present can alter our past as completely as they can alter our future.

This throws our personal reality as a fixed, stable configuration under a very large bus. We assume a hell of a lot about who we are, and by extension, who everyone else is as well. Until we step forward onto that bridge of understanding between the worlds, we are as utterly phony as the imposter in the film *Catch Me If You Can*. Running yourself to earth is every bit as challenging as catching the movie fugitive who has as many identities as Lon Chaney (you may have to look him up) or Martin Short.

Now, if you aren't as real as you think you are, how real can other people you encounter in a holographic reality be? After all, you can't be sure of the degree of your own reality, so by what standard are you evaluating the reality of everyone else? Let's add a further complication to the puzzle by examining some weirdness among other people who think they are other people. I am referring to people who have multiple personality disorder.

According to an online article appearing in the Science section of *The New York Times* for June 28, 1988, by Daniel Goleman,

> In people with multiple personalities, there is a strong psychological separation between each sub-personality; each will have his own name and age, and often some specific memories and abilities. Frequently, for example, personalities will differ in handwriting, artistic talent or even in knowledge of foreign languages (par. 7).

Again, you have the emergence of several different entities, each with his or her own history but within what appears to be a host personality. Remember the sixty-million-year-old rock you collapsed out of quantum wave function despite the fact that the rock didn't exist until you observed it? I suspect that something like that is going on with people who have multiple personalities. Are they created by the host personality's observation of them? Who exactly is the observer? Are all of the personalities observers by turns? Whatever the case, they are as "real" as anybody who doesn't have multiple versions of him- or herself in play.

Dr. Frank Putnam, a psychiatrist at the Laboratory of Developmental Psychology at the National Institute of Health, contends, "A given state of consciousness has its biological reality ..." (par. 8). Think about that for a few minutes. What is Putnam actually saying? I think he is saying that consciousness creates physical reality. Documentation of biological changes in people with multiple personality disorder has been going on for more than a century. According to Goleman, some of these changes include "the abrupt appearance and disappearance of rashes, welts, scars and other tissue wounds; switches in handwriting and handedness; epilepsy, allergies and color blindness that strike only when a given personality is in control of the body" (par. 9). Katy, bar the door! Wait a minute, you *are* Katy, aren't you?

If each personality, through the exercise of its consciousness, is creating its own biological reality, isn't it as real as the host personality that is creating *its* own biological reality? How many fields of consciousness are in play here? Each would seem to be capable of collapsing wave function so as to create its own reality. Or does the host consciousness split itself into sub-fields of consciousness and selectively collapse wave function through different personalities? This would be a very cool ability to have. As it is, we feel stuck with one observer who calls all the shots, leaving us with a single reality, which isn't always that pleasant. By *playing like* we are a different personality, could we shift observers so as to create an alternative reality? Could we believe that completely? Could we effectively become another personality entirely, erasing temporarily the personality we were previously? How would we reconnect with our host personality when we had no recollection of who that personality was? Intriguing stuff.

So don't be hasty in assuming that the host personality is the real one and that all the others are fakes. The host personality is simply the one that conforms to consensus reality expectations, and we know what that's worth. What the researchers are at now is discovering the mechanism that enables consciousness to change its biological reality so that it can create "good" qualities instead of "bad" ones. The goal is to be able to manipulate consciousness for the better, but who is manipulating the manipulators? Can consciousness be segmented into an inside, authentic pattern in control of a false outside one—the consciousness of the therapist manipulating the consciousness of multiple personalities?

I suspect a certain amount of holographic tyranny is involved here. "You do want to be normal, don't you?" asks the therapist, who wants to constrain and control an aberration in consensus reality behavior and then use the mechanism discovered to advance the morphic field of consensus "normalcy." What the therapist wants to do is "integrate" the spurious personalities back into the dominant one to restore a whole, normally functioning personality.

It's like Jill Bolte Taylor's restoration of her left-brain function. She wasn't who she believed herself to be while her right brain had control of her awareness. She discovered that there were aspects of herself that she really liked that had been hidden behind the dominance of her left-brain yadda-yadda. Nonetheless, she couldn't function in the consensus reality surrounding her using only her right brain. Talk about being between the worlds! But Jill had a malfunction, which existed as a malfunction in holographic reality. She had to correct the malfunction to bring it into line with the preexisting consensus morphic field. What if she had decided to correct the morphic field instead and bring it into line with the observations of her right-brain awareness? No go. Persistence won out. A reality in the hand is worth two in *potentia*, it seems.

Let's enlarge the context. Over the years, people who have had near-death experiences report that physical disabilities they had while physically alive vanished during the interval in which they were clinically dead. I recall one case in particular in which a woman who had been blind from birth experienced perfect sight after she had flat-lined in the hospital. Her amazement was, to say the least, considerable. After she had been resuscitated, she was again totally blind. Sounds like she shed one reality for another by shifting her consciousness, which raises the question of how implacable her established consciousness really was. It makes death sound more like one of those reversals of conscious orientation—discovering that what you thought to be real is the opposite of what actually exists—than a termination of being. The difference in the case of death is that, instead of changing one biological reality for another as in the case of multiple personality disorder, you are changing a biological reality for a non-biological one by experiencing a deeper shift in consciousness than occurs with a holographic-level personality shift.

You see how tangled all of this reality of other people is. It isn't limited just to who is real in our holographic consensus reality, but it extends to who is real among multiple versions of a shifting personality that seems to defy that reality, to those who are part of that reality, but seriously doubt it, and even to those who have physically left this reality and crossed over into a separate one altogether. As you read these categories, you are assuming your own reality to be the definitive one, but with all these other versions in play—all depending upon differing states of consciousness—you are mad to do so.

And yet—individuality prevails. Ready for another paradox? In the Jungian view of human awareness, the process of individuation is considered a move toward wholeness. Aren't the two mutually exclusive? How does that work? Paradoxically, it is our acceptance of paradoxes that enables us to become integrated, whole beings. The Jungian process of individuation involves a progressive integration of seeming opposites in order to complete one's realization of participation in a universal whole. That's why Carl Jung became so interested in alchemy. The art and science of alchemy involved the "marriage" of the male and female components of the human psyche in order to create the hermaphroditic superior being. It is the power of three, two opposite polarities combined produce a third transcendent power.

Although alchemy has a long and distinguished history in both Western and Asian culture and has attracted some of the most brilliant minds of the Middle Ages through the Renaissance (even Isaac Newton, in the seventeenth and eighteenth centuries, was an alchemist), Westerners still balk at accepting paradoxical thinking. This peculiarity of Western thought makes disobliging reality a particularly difficult enterprise.

A rather perceptive fellow named Dirk Gillabel (not an academic authority on Jungian psychology, but he seems to know his stuff), author of the Internet article, "The Individuation Process," notes that, to become whole, to complete the individuation process, "it is necessary to accept both the superior and the inferior, the rational and the irrational, the order and the chaos, light and darkness, yin and yang" (The Experience of the Self", par. 1). So also must we accept the discovery of our cosmic identity by defining our individual one. The individuation process, Dirk says, "is one during which one integrates those contents of the psyche that have the ability to become conscious. It is a search for totality. It is an experience that could be formulated as the discovery of the divine in yourself, or the discovery of the totality of your Self" ("Individuation", par. 1). From individuation, one moves to the second phase, which Jung called the transcendental function. "This function," Dirk adds, "has the capacity to unify the opposite tendencies of the personality," leading to the "goal of transcendence ..." ("Transcendence", par. 2).

Thus, by becoming more completely an individual, we transcend our individuality. This further complicates the determination as to whether other people are real. Everyone is committed, to a greater or lesser degree, to completing the individuation process. So there would be a question as to how far along the individuation process each person has progressed. From the Jungian viewpoint, the less a person has individuated him- or herself, the less "real" he or she would be. This pulls in the whole levels-of-meaning aspect of consciousness, plus stepping forward onto the bridge of understanding.

In other words, there is a point in the individuation process where, as one finds more and more of the divine aspect of oneself, one shares the bridge of understanding with others who also find themselves between the worlds. One world is the world of individuation and the other world that of transcendence. There is the singular, which encompasses the totality. Individuation begins with disobliging reality—neutralizing the persona, integrating the shadow—and proceeds to realization of a separate reality, which leads to transcendence.

What is real about other people is the degree of their inclination (their angle of momentum) to embrace the paradoxes needed to accept the total reality of their true selves. Until they begin that process, they are like the unformed wave function of infinite possibilities—potential locations in a probability cloud, extras on a film set, part of the scenery—not nonexistent, but not completely real either, at least not real in the way we understand reality—in short, a paradox.

11

Death.
An Exciting Change of Pace?

I recently read William Peter Blatty's book, *Finding Peter: A True Story of the Hand of Providence and Evidence of Life after Death.* (Okay, get a grip. You got through exposure to the Bizarro world. Suddenly you're worried that things are going to get spooky. Relax. They've been spooky all along.) About halfway through the book, Blatty says this:

> It was the renowned British physicist Sir James Jeans who, after years of attempting to unify all of the laws of matter in a single coherent theory, at the last decided that the only way in which that could be done was by assuming the material universe and everything in it were thoughts in the mind of God, while in the decades since then the quantum physicists have been telling us that there are no such things as things, there are only processes, and that matter is a kind of illusion. We live in a universe, they say, in which electrons can travel from a point in space to another point millions of miles away without traversing the space in between; that a positron is an electron that is traveling backwards in time; that in any two-particle system, changing the charge of one of the particles instantly changes the charge of the other even if the particles

are several light-years apart; and that nothing really exists until it is observed. And so I ask: In a universe such as this, should there really be any such thing as surprise? (137).

Ah, a poignant question. Blatty is casting grave doubt upon reality as we are taught to know it. He is laying the groundwork for the disobliging process. His motive is a powerful one—the loss of a loved one. If one accepts the reality we are handed from childhood, then when someone one knows well and cares about physically disappears, he or she must no longer exist except perhaps, if you are of a religious persuasion, as a disembodied spirit in a faraway location watched over by angelic custodians. It's the total bye-bye.

In the classically Newtonian universe we favor, matter walks heavily and carries a really big stick. If you can't hold it in your hand, then it must not exist. How can someone you've known and touched and interacted with for years suddenly not be here anymore? Where did he or she go? Was he or she ever really here in the first place? If you loved someone, then the questions take on an agonizing insistence.

For Blatty, as for many of those, maybe even most, who have experienced the physical disappearance of a dear companion, these questions initiate an obsessive quest for answers but answers that do not include, "Well, he's just dead, and that means he doesn't exist anymore." It's the passionate rejection of that answer that keeps the quest alive. The easy dismissal is that "it's just denial," or "some people just can't face the truth." In other words, people are so scared of having to die, to run the risk of becoming nothing, that they refuse to believe in the death of anybody else. We all want to live forever, they say, but life takes a lot of forms that some would be more than willing to do without. Ask those who have attempted suicide.

So why the ongoing presumption that there is more to life than what is experienced in physical reality? It's as persistent as the reality itself. Hey, that could be a clue. It's the balancing out of excessive potential. The more potential that is loaded onto physical existence through focused attention, the more the potential for nonphysical existence is increased. Why sure, it's the old yin-yang. The paradox.

Death is the fulcrum, not the heavy end. It's the gap between on and off. The balance is not between life and death; it's between life and afterlife.

The opposite of something isn't nothing. That's the old dualistic confusion. The opposite of something is anti-something, which is still something but with a negative charge. Or the opposite of something could be after-something. It's just as valid, just as good, or in this case maybe better. We know this in the deepest core of our being. That's why we need to keep heckling the sly illusionist of the here and now. Blatty calls death "that liar and fraud." He's royally pissed because his son Peter has vanished but even more pissed because he had been led to believe that his vanishing meant something very specific and unalterable that is palpably untrue—that to disappear physically is to cease to exist at all, or at least in any clearly interactive way with physical reality.

That's why it's called *physical* reality, to distinguish it from any kind of nonphysical reality, which by definition has to be completely conjectural. The issue here is most people's unwillingness to loosen their grasp on physical reality so as to allow for an experience of nonphysical reality. It's the fear factor once more. The paradox here is delicious. Physical reality is something to be feared because it is inherently dangerous, but at the same time, nonphysical reality is to be equally feared because it isn't physical reality.

For Blatty, something of physical reality (his son), which he dearly loved, passed into a nonphysical state, causing a kind of splitting of his consciousness and emotional equilibrium. Now he finds himself walking between the worlds, his physical self still involved in keeping on with the necessary activities of a physical existence despite their now irrelevant triviality, while at the same time his deepest emotional attachment, the very heart of his being, has been relocated to a nonphysical dimension that he knows little or nothing about. He is conflicted on a scale that is staggeringly incomprehensible. Through no fault of his own, he has been subjected to a state of existential schizophrenia so horribly overwhelming that no avenue of escape can even be contemplated except that of terminating the consciousness that allows this unendurable conflict to continue. In Blatty's case, his devout commitment to his Catholic faith gives him a meaningful context for sorting out his chaotic anguish and provides a means for congregational rescue.

This kind of rescue, which depends upon a specific morphic field, is not to everyone's taste and imposes certain restrictions on how the rescue is carried out, namely in terms of the revealed dogma of a divine authority whose own sacrifice obligates its benefactors to adopt a posture of worshipful gratitude. This, in my view, involves too much stage dressing. Assuming that Blatty was not speaking tongue-in-cheek, he accepts the basic theories of quantum physics, the critical one being "nothing really exists until it is observed." That includes Lebanese Catholicism. By drawing on the faith and teachings of his church to carry the force of his dilemma to an acceptable outcome, Blatty is putting the cart before the horse. The horse is the observer's collapse of the wave function. Lebanese Catholicism is only one of an infinite range of possibilities that can serve as a cart.

Blatty's religious preference is an arbitrary one based on his past experience and the involvement of people who shaped his personality and character. If we are mired in an illusory reality, of which consensus religious faith is an expression, then perhaps we ought to cut out the middle man and deal directly with the source of the illusory effect by disobliging the reality that creates it. That includes whatever religious institution may be masking the illusory reality. Religion is supposed to be about spiritual reality, about the separate reality of the implicate order, but it too often is couched in terms of dogmatic beliefs and prescribed behavior within the context of the consensus reality. Only the mystics escape conformity to the established protocols and most often by being allowed their subjective visions as a symptom of a divine madness tolerated so long as no fundamental departure from the essential god-figure template occurs.

Consider that institutional religion, which is almost always some version of a revealed religion (revealed through inspired gospels, sacred texts, prophets, avatars, incarnations of the divine, etc.), confirms the idea that there are two separate realities, the physical or worldly one and another nonphysical, unworldly one. The popularity of the religion depends heavily upon this distinction. This means that the agents or priests of the religion are needed to act as translators, interpreters, and intermediaries between the physical world and the nonphysical one. They are guides and authorizers who help the faithful find their way to "salvation" in whatever form that may take.

The key difference here, though, between our implicate and explicate orders, between consensus reality and a separate reality and the this-world/other-world of popular religion, is that the religious view rigorously maintains the opposition between the two polarities. Even though physical reality may have been redeemed in some fashion, it is still opposed to the heavenly alternative. The lines are clear. "This" is not "that" and can never be. The devout must deny "this" in order to obtain "that." There is no consideration of the possibility that "this" is a mistaken version of "that." To influence its adherents sufficiently, an institutionalized religion must ironically endorse the reality of an imperfect, potentially evil physical existence in order to maintain an adversary it can oppose, thus giving it a noble mission and a cause worth supporting by people in general.

Despite all this intricate elaboration of "right paths" and "noble truths," "venial" and "mortal" transgressions, "indulgences" and "absolutions," "dharmas" and "karmas," "Salat" and "Hajj," "Shacharit" and "Shabbat," etc., *ad infinitum,* the most basic, in-your-face stipulation of this universe is that "nothing really exists until it is observed." That's how things work in *this* holographic projection anyway. If Mr. Blatty took it from this point of view, he would develop a very different approach to finding his son, who really didn't go anywhere because there is no "where" to go to. Peter merely got free of his holographic encumbrances. Unfortunately for his dad, this liberation included Peter giving up all sensory interaction and communication with his previous holographic dimension.

This, of course, is a broad generalization. Remember that this consensus reality is inconsistent and often contradictory, allowing for all kinds of exceptions to its standard of normalcy, which is conveniently called the "paranormal" in order to marginalize it and make it sound highly questionable, if not totally irrelevant or outright insane. So contact between "this" and "that" occurs, and it occurs all the time. It just doesn't make the news unless it can be presented as a random curiosity in an otherwise safely comprehensible world.

Death is actually a staged performance with all its props, costumes, and makeup provided by the prevailing consensus reality. Thus, its catastrophic impact on everyone affected by it has to be assimilated into that reality in a way consistent with its stage-show nature. Its implications are part of its script, and we enter the performance, acting out our roles as bereaved mourners and emotionally devastated participants in a well-rehearsed scene. The scene is so compelling, so dramatically absorbing, that its reality is unquestioned, and its raw persistence allows its intrusion into areas that are beyond its scope. We are so captivated by its performance that we allow it to shape our perception of what transcends it, of what makes its hackneyed tableau even possible.

All the stage dressing of death seeps into whatever we imagine must follow it. It's like a mediocre stage magician who, once he has performed his electrifying act on stage, follows you home to your personal reality and continues to perform clumsy, unconvincing tricks throughout the rest of your life, expecting continual applause and curtain calls. He just doesn't know when his show is over. He won't acknowledge the fact that he has left the proper venue for his act, which is a small stage in a poorly lighted and shabby theater with unmarked exits. He becomes both a posturing, self-important bore and an unendurable boor.

Because consensus reality has managed most of our lives from our childhoods to our deaths, it pretends to continue its management after we no longer have any use for it. It does this by intervention into the grief of those left behind by those who have entered into a different and more expanded realm of existence. For fear that the mourners may begin to suspect that there is something beyond death that invalidates the reality of the life that precedes it, the advocates for the prevailing reality begin trying to return the bereaved to its reassuring grasp.

It's all about limitations and restrictions, coming to terms with restraints as an inherent condition of existence. The bereaved are encouraged to "get on" with their lives even though there has been a pretty dramatic interruption of what those lives seemed to be about. "Don't worry," they are advised, "things will get better. You'll get over it." In other words, go back to believing in the popular illusion of reality, and you'll eventually forget that any other possibility ever existed. Rather than being a wake-up call, death becomes the main incentive for committing oneself even more completely—even avidly—to the illusion that hides it. "Try to be happy," the sorrowful are urged. "Try to keep busy." What is there to be busy with? Why, reality, of course. Normal living. Everything that death isn't about.

Death is an aspect of "that," the portal to an alternate reality. "This"—the real world—is the opposite of "that." People who accept that into this are considered at least morbid if not hopelessly lost depressives. "Get out more," "take up a hobby," "go on a trip." Otherwise, what? Otherwise more and more of "that" will creep into your life, and "this" will begin to pale by comparison. Triviality and small talk will no longer be sufficient. The most mundane of things and activities will become pregnant with metaphorical implications. Every breath you take will become an opening into unimaginable possibilities.

And what of the departed beloved? Did he or she really depart, or did he or she assume an unfamiliar perspective? Does the departed one have to wait out the demands of a keyhole reality in which those left behind dither with busywork so as to "get back to normal"? To savor the ongoing presence of the departed, to feel again the touch of his or her hands, the sound of his or her voice, the unmistakable phrasing of his or her most cherished ideas and the looks of his or her happiest expressions and endearing moods and attitudes—these are not just memories. They are facets of an interactive reality we are taught to put on hold until we are through with this one. But they are one and the same.

It is never a choice between this and that. It is always both this *and* that. And if you can't be happy apart from your beloved, you are not to be pitied and rehabilitated. You are just lovesick in a very beautiful way. You have seen through the clumsy illusions of a divisive reality and can no longer feed its divisiveness with your willing acceptance. You are now connected to "that" in a very personal and compelling way. You walk between the worlds without being completely taken by either. What is mistaken for unhappiness at being left behind in this world is merely the tension of a love drawn taut between two seemingly different aspects of a totality that ultimately is one. What appears to be sadness is the emotional knowledge of this unifying potential. It waits upon the opening of totality's least action pathway, leading every abandoned heart back to its beloved.

12

This *and* That

As long as I have once again made reference to this and that, I might as well elaborate on it a bit more to make my meaning, if not clear, then at least slightly less murky. We could consider this the logistics of this and that. Early on in this exploration, I had our reality doubter seizing the throat of a reality believer and demanding, "Why this instead of that?"

In effect, the doubter is asking why we must experience this consensus physical reality to the exclusion of the nonphysical separate reality that underlies it. Why must we settle for the explicate order as the only version of reality we can know instead of having the choice to experience the implicate order as well? After all, they are described as being interdependent. Shouldn't we be allowed to investigate the reality of this connection? Why are we restricted to a three-dimensional reality when there is ample evidence that we exist in many more dimensions than that? Why are we limited to a linear and sequential left-brain reality as our standard of what is popularly sanctioned authenticity instead of being allowed significant bleed-through from our non-linear instantaneous right-brain reality to offset the mechanical, cause-and-effect structure of the former? That is always present, but it is obscured by the persistence of this, which is aided and abetted by the consensus majority, which prefers this because it is frankly shallow and obvious and literal.

Seems like a clear choice, doesn't it? If only it were so. Say you have a madcap moment and decide to choose that over this. You have decided that this, from a value-fulfillment perspective, is bad and that that is good. Now you're making a judgment based on dualistic thinking, which is typical of this reality, and so you are strengthening the persistence of this over that and achieving the opposite of what you intend. The catch is that what you consider that is equally this from that's point of view, which is holistic and inclusive. This, on the other hand, is perfectly happy to exclude that as any kind of legitimate reality because it prefers to entertain opposite, exclusive polarities as the basis of its universe of meaning. The point is that there is no choice between this and that. Life is made up of both this *and* that simultaneously. We have just been trained not to see things this way. So now, if you want more of that, you don't get it by eliminating this to make room for it. You need to find that in this by altering your consciousness from the belief that the two are opposed to each other. This involves sensitizing your awareness to the presence of that in every moment of this. From this new perspective, you become able to appreciate the fact that this is really just another version of that. This becomes that with the realization that this can never exclude that. Now what?

Let's assume that practically everything you have known and experienced is some version of this. How do you experience that? Remember that reality is a feeling. What does this feel like to you? Notice especially if you feel that this is not you, that it is outside of you, and that this causes things to happen *to* you, whether they be good or bad. Is there a feeling of separation between yourself and this? That feeling of being different from this is the feeling of its reality.

So what is the feeling of that? That is the feeling of who you truly are when you're not being overcome by this. In other words, when you are being truly who you are—and you know that comfortable, relaxed feeling of being in tune with yourself, with having the space to let yourself step forward into full expression with delight—then you are in touch with that. Find the frequency of that feeling. Tune it in as you would tune in a radio station broadcast frequency. Get the strongest signal you can. Notice how different that feels from this. That's because the feeling of that goes beyond your immediate perception.

There are deeper levels to being who you are that go beyond the capacity of this.

Don't fall into the trap of thinking the real you can be described in this-reality terms. It has nothing to do with how old you are, where you come from, your gender, your race or ethnicity, what kind of work you do, or where you expect to be in five years. You already know what it feels like to have someone else try to make you someone other than who you deeply feel you are. That deeply felt you is the one you are looking for in order to find the frequency of that. It's your heart, the heart of you, the soul of who you know yourself to be.

Be careful here. You can fool yourself into believing you have found your frequency while you are still allowing someone else to make you into a person you are not. That someone else who is automatically exempted from exerting such pressure is the deity you worship. As "the deity," he, she, or it is credited with being ultimate reality itself and therefore is beyond your puny, imperfect assessment of its role in making you someone other than you truly are. But gods and goddesses are created by our consciousness to serve questionable purposes hidden from our own awareness. They can be used to provide the worshiper with a sense of moral superiority that can become highly judgmental and further the imposition of dualistic thinking on your personal reality. A deity can provide excuses for all kinds of personal excesses that are driven by guilt, fear, revenge, or megalomania.

A deity can take many forms. It is basically a focal point for some type of unassailable authority that transcends the reality of this precisely because it represents that. Paradoxically, nonbelief can assume the function of a deity. Atheism is just as theologically assertive in a negative way as its conventionally religious counterpart is in a positive way. Whatever controls your thoughts and feelings in an authoritative way, whether it be religious, philosophical, political, cultural, sociological, etc., it functions as your deity and suppresses your freedom to be who you truly are. Remember that uncertainty should be the defining temper of your disobliged reality. The Zen tradition speaks of "putting another head on top of your own," which refers to imposing an outside authority or conditional framework on your own spontaneous insight. Usually it is some fabrication of left-brain origin that tries to usurp your authentic autonomy.

Get over it. Being who you truly are is a that frequency that you'll want to anchor into this reality, thus experiencing more of this *and* that as complementary aspects of an undivided whole. This unity alone will do more to disoblige reality than any contrived strategy itself can achieve.

You can still perceive the operation of your double-slit awareness experiment in the merging of this and that. You are merging single-slit particle reality with double-slit interference pattern reality as aspects of the same experiment. You learn how to observe one and not observe the other, getting two seemingly different results from the very same process of shooting particles at two slits in a screen. You have this *and* that, and it is only your conscious attention or the absence of it that either separates them or blurs them together as a probability cloud. Whenever life (holographic reality) starts bunching up on you and seems too much to bear, remind yourself that you are in a holo-deck of your own choosing where you can stop and in a commanding voice, say, "Computer, end program!" You walk between the worlds.

13

Multiple, Probable, Loveable You's

What I am going to touch on now has to do with who you truly are and why the discovery of that frequency involves you in an open-ended interaction with the reality of that. This is a subject that can be quite complicated and extensive, but I want to simplify it as much as possible without distorting or falsifying its authenticity. Many sources are available to you for deeper study of the implications of recognizing your true self, so I won't try to repeat more than a brief summary of what they say here.

Remember the probability cloud of the possible locations of an atom that has not yet been observed? You are like that atom. Before your observation of a specific location for yourself, you exist as a cloud of infinite probable selves. All these selves exist in both your past and your future from our time-oriented perspective. They are all of the probable you's that were created when you had the intention of making a choice. Once your choice is made, the remaining probable you's don't disappear. They continue to pursue their own courses of action in other probable universes. They follow all of the variations possible from your choice of a particular course of action. Because, says Norman Friedman, "we are serially and three-dimensionally oriented," we are prevented from "perceiving our existence in a sea of probable events" (135). He goes on to quote Seth on the subject: "It is impossible to separate one physical event from the probable events, for these are all dimensions of one action" (qtd. in Friedman 135). Seth adds, "All of the probable events of your life exist at once.... Since your activities physically must be fitted into a space-time framework, only a minimum of those probable events will physically occur" (qtd. in Friedman 135).

In other words, since we are a part of this, we cannot experience physically those events that remain part of that. Yet they do exist and incredibly, influence our probable physical events now from both our past and our future. Friedman says that "All probable paths," according to Richard Feynman, "are aspects of the actual path taken" (136).

If you think that's a *wow* moment of realization, get this statement from Friedman:

> We are secure because we are choosing from unpre-dictable fields of actuality those that suit our own par-ticular nature. The ego, or sense of selfhood, jumps in leapfrog fashion over events it does not wish to ac-tualize, leaving other portions of ourselves to choose the events we have *not* chosen" (136).

Double wow! Add to this the following quote by Seth from Jane Robert's *The "Unknown" Reality: Part 1*:

> Your desires go out from you in time, but in all direc-tions. On the one hand as a species your present forms your future, but in even deeper terms your pre-cognitive awareness of your own possibilities from the future helps to form the present that will then make that probable future your reality (Session #690 March 21, 1974, 140).

Katy, bar that door, and then bar it again! Do you begin to get the immense significance of what it can mean to find the frequency of your true self? That self is not an isolated fragment of consciousness having a separate awareness unconnected to both this and that. It exists throughout both the explicate and implicate orders both inside and outside of space-time. This is a huge realization to take in. The identity of your true self incorporates (a somewhat misleading term because we are not confining our discussion exclusively to the cor-poreal) all of your probable selves existing in other universes in both your past and your future. Your true self is being informed by all of those probable selves but seldom in a conscious way. Now that you know this is going on, you can begin to develop an awareness of this enormous process. You no longer have to use your limited present to try to figure out what the future holds for you or recall your past to understand how it is affecting your present. Your true self is actively participating in both time frames simultaneously. By not isolating your sense of self in the present of this reality, you can entertain or play host to the experience and wisdom of all of your probable selves who exist in multiple universes in that reality. You just need to find the frequency of that total self.

It may have occurred to you that having an infinite number of probable selves can create some confusion about who is the *real* you. All your other you's are busily collapsing wave function in their realities and creating more "sub-you's" to deal with. How can you be sure you are not a probable you who was created by the real you, with whom you have never had conscious contact? Remember that you are the one who acted as the observer in creating all of your probable other selves. You are the one who got the ball rolling. As a result of your initial collapse of the wave function when you chose a course of action, you in effect cloned yourself into multiple parallel existences. You are still the originator of the process, so you are still who you have assumed yourself to be from the beginning. The difference is that in understanding the immense implications of who you really are, you just got a whole lot bigger

As previously stated, "All probable paths are aspects of the actual path taken," according to Feynman. All probable you's are aspects of the actual you whom you believe yourself to be. Understanding that insight puts you in touch with both this *and* that and goes a hell of a long way toward disobliging your reality. The important thing is to *find the feeling*. Try to locate it spatially in your immediate surroundings. Is it over your head, behind your left shoulder, or in front of your solar plexus? Remember that space is not empty nothingness. It has its own geometry, a structure that you can explore. Let it show you where your true self resides, not as a specific physical point but as a concentration of feeling. Once you locate that resonant area of feeling, you can return to it whenever necessary. You can make that frequency an aspect of this while it remains an aspect of that.

We are expected to play the game. It's sort of like playing three-dimensional chess as it appears in *Star Trek*. However, in this version of chess, it is played in far more dimensions than those inherent in *Star Trek's* seven-level version. Space-time is the chessboard with its black and white pattern of squares. This limits us to a dualistic playing area that requires either/or choices. Some levels are fixed, and some can be moved. This enables the player to move both his or her individual pieces and the levels of play. The implication of this arrangement for my purposes is that a player can change his or her circumstances by changing the position of the pieces and alter the strategic context of those circumstances as well by raising or lowering the level on which the play is conducted.

This pretty well describes our situation in physical reality. We're stuck with the chessboard, the pieces, and the rules, but we can claim more initiative for ourselves by changing the level on which the game is played. To the casual observer, the play remains unchanged, but for the player, much more is going on than can be observed by a spectator.

You see the correspondence here between disobliging reality and playing multilevel chess. The chessboard, the pieces, and the rules are consensus reality. The levels represent the levels of consciousness on which the game can be played. When a player changes a level, he or she changes not only the significance but the actual physical circumstances of the game. The chessboard morphs into a holographic field, and the pieces become contact points between two worlds. Play no longer amounts to moving discreet pieces from fixed location to fixed location but allows the pieces to become bridges between fluctuating fields of information and meaning. The overall purpose of the game is no longer merely to defeat an opponent but to gather more and more awareness by turning opposition (the opponent's pieces) into the expansion of possibilities (an ever-expanding field of play).

This is not an easy game. The loss of a piece has greater consequences than in traditional chess because, as a bridge to more information, the loss of the three-dimensional piece lowers the player's capacity for "reading"—that is, deriving meaning from his or her situation vis-à-vis the holographic playing field. As the loss of pieces mounts, the player will be forced to play on a lower level without being able to improve his or her situation. Taking" the opponent's pieces (which represents gaining awareness from opposing circumstances or events by internalizing their positive potential) is the prime directive of the game. The pieces, like those in traditional chess, have different values. Loss of lower-value pieces will not change your level of play. Capture of higher-value pieces from your opponent will raise your level of play. The relative levels of play between the two players will affect the values of the individual pieces. Pieces played on a higher level have more positive potential than pieces played on a lower level.

This has been a much more extensive analogy than I had originally conceived. I think it puts disobliging reality into a clearer perspective, however. Ultimately, the reason for disobliging reality is to enable you to play the game of this *and* that at increasingly higher levels of awareness. The final payoff is to transcend the game itself. It is a wickedly clever game in that it conceals the fact that it *is* a game until a player happens to discover elements of game theory in its pretense at being simple reality.

Einstein, a very clever and perceptive explorer of the levels between the worlds, was onto this from the start. He advised, "You have to learn the rules of the game. And then you have to play better than anyone else." (This very popular quote is attributed to Einstein, although it is disputed. Since the past is just as unreal as both the present and the future, I declare, here and now, that Einstein made this comment during a night of carousing in a Stuttgart *bierstube* in 1926. He was referring to the game of *Arschgrabschen,* which can certainly be played on different levels.)

14

The Game Is Afoot

Now let's see what sort of context for your awareness we have at this point. There is the double-slit experiment, which gives you access to both the explicate and implicate orders, the unreality of space and time, the flickering off and on of the universe, the collapse of infinite possibilities into a holographic semblance of a solid object or event, the absence of cause and effect, the uncertainty of everything you perceive, and most recently, the creation of a cloud of probable you's every time you choose a path of action. Finally, the whole mess becomes the foundation of a game you must play that has something to do with expanding your awareness. Why in hell would you get your knickers in a twist over such a wispy, ephemeral "reality"? Once more, it's because of "this" reality's persistence. You know how involved you can get in a game even though you realize it's only a game? It's the willing suspension of disbelief. We actually *want* to suspend our belief that we are only playing a game. We don't want to *disbelieve* in the reality of the game. From a higher perspective, we act like moths drawn to a flame. We *need* to act self-destructively by running headlong into stone walls, by creating all kinds of limitations for ourselves and then spending our lives trying to overcome them (Ramana Maharshi has noted this curious aspect of human behavior with, I suspect, wry humor.)

Here is the bitterest pill to swallow, which is still the operating principle of all aspects of our universe: nothing really exists until it is observed. The corollary to that is even more disturbing: nothing is real; everything is made up. You would think that the choice between struggle, pain, and suffering and the conviction that nothing is real would be a slam-dunk. But it never is. No matter how many levels of meaning we manage to ascend, there is always something that catches our attention, something we fear or that we are curious about. Then off we go, slamming into stationary objects or trying to fix, change, or improve the things we create so as to eliminate their flaws and imperfections.

Some theorists think that realizing the absurdity of the whole enterprise would make the experience of game-playing fun and entertaining. In other words, anyone with an enlightened view would be drawn into participation in the game for the sheer enjoyment of it. This is bullshit. No matter how much of that you are able to anchor into this, it's still going to hurt, and often hurt terribly. If that's part of the fun, I'll take vanilla.

We have seen "highly evolved" people, often gurus of one sort or another, whose daily lives are awash with bliss and love. Many of us aspire to that state, thinking that level of joyous detachment will guard us from loss and disappointment, struggle and loneliness. These spiritual overachievers seem to have found the way to live unaffected by any of this because they are so totally grounded in that.

Realize that that assumption has a dualistic basis. The actuality is that this *is* that. Being wholly absorbed in that doesn't eliminate this. It's a package deal. Nowadays it's called "bundling." The programming that comes to you is preselected. You don't have the option of choosing your package, but you do have the ability to select the channels you watch. Still, you are compelled to make your choices using a television set and a signal that you pay for. Let's say that your particular TV set is equipped to give you 3-D reception. There you go. You're experiencing a holographic reality. There are certain inherent conditions that you must accept with this technology. In terms of the analogy I am using, the technology of this reality includes pain, suffering, and death. Yet you can't resist switching on this damned TV because the reception from that is too poor.

There has to be a point to this arrangement, and it is a paradox (are you surprised?). There are stakes involved. Discovering what I have been presenting to you requires some compelling curiosity on your part and the tenacity to keep digging into it no matter how much difficulty you encounter. The payoff would seem to be transcending all the limitations imposed by yourself onto a seemingly physical reality, yet death and loss are never escaped. The individuation process itself is no romp through the garden of Eden either. Losing comforting illusions can be utterly vitrifying (the production of molten glass by extremely high heat like that instantaneously generated by a mega-max explosion; "devastating" has been done to death).

So transfiguration or apotheosis is not in the cards for us right now. The point is that we have to make the most of the options we are given within some basic parameters. This is not to say that those parameters will never be transcended. Bypassing physical death and the emotional trauma of great loss could be an eventual adaptation of our species to a quantum reality. Travel between dimensions may one day be accomplished without the dissolution of the body. At this time, death is just a crude method of recalibrating our oscillation frequency. I feel it is likely that we will outgrow our need for a holographic body oscillating at a specific frequency and develop a form that can change frequencies to suit the dimension we choose to visit. Once outside of a holographic reality created by our consciousness, we may never be subjected to agonizing loss because the limiting conditions for that experience may no longer be in play.

The overall point of the current arrangement seems to be the development of what used to be called character. Probably the most valued traits of character are courage, independence, self-reliance, integrity, and perseverance. To these I would add a sense of humor, which serves the purpose of keeping the entire constellation of character traits and the challenge of playing the game in perspective. We must deal with all of the seeming random events of physical existence, both joyful and crushing, and adapt our inner equilibrium to their configuring effects.

Say that you have experienced a staggering loss. Once you re-covered from this blow, you undergo another brutal defeat. Common sense tells us that this sequence is not just unfair but outrageously intolerable. Why? Because it violates a proper sense of proportion. There is no balance to these catastrophes. They are monstrously excessive. Yet why should these events conform to our expectations of what is acceptable? We have no hand in arranging the larger con-tours of our physical experiences once we have begun to play the game. Otherwise we would choose immortality that is free of pain and suffering. We do have an amazing amount of discretion in shaping the quality of what we do experience but apparently not of restructuring the basic framework in which the experiences take place. The game is afoot, and we do the best we can to play it well.

15

Aiming for the State of Grace

Without the knowledge and understanding of that being an aspect of this, we are born to lose. In other words, we remain rank beginners in the game. Here's where things get subtle. You have developed an understanding of all of the aspects of experiencing a reality that is a complete illusion you have created. You are walking between the worlds. You have internalized a mood of uncertainty about what appears to be going on in the world. You recognize that you are living a consensus reality that has been prepared for you by an enormous and very powerful morphic field to which you are contributing your attention, emotion, and conceptual validation. How is that supposed to feel?

Practically everyone has at one time or another experienced periods of guidance or inspiration. Artists, musicians, writers, athletes, and just about all people who have engaged in a creative activity (including the most seemingly mundane like cooking, gardening, hiking, etc.) have had a moment when it seemed another awareness had overridden their own. Suddenly, it seemed as though another consciousness had taken over the activity they had been engaged in and become the dominant mode of perception. The feeling is of being a conduit for something outside oneself. Whatever the activity, the sense is that it's being done by someone or something other than oneself—something much greater than oneself. The result is a feeling that whatever was accomplished—a painting, a poem, a spectacular play in football, a choice of arrangement for a garden, etc.—it was beyond the normal capacity of the person doing it. It was accomplished through a state of grace when one was carried along by an intuitive awareness that knows far more than ordinary perception can supply. It was a direct experience of that. It was a perception of that in this.

Unfortunately, we are led to believe otherwise. We have invented "coincidence" to allow for all of the things that occur outside of our "normal" abilities. We believe these things are the result of luck, a fluke, or being in the zone. But Asian culture, to which I referred to earlier, allows for the play of yin in its reality. This yin component has been actively sought for through numerous traditions in Asian history. It has been especially prominent in the practice of Asian martial arts. Zen archery seeks to achieve this spontaneous state of grace through the actions of drawing a bow and aiming an arrow at a target. The release of the arrow is dependent upon the archer's recognition of the state of grace in his or her awareness. In effect, the arrow is shot by a mysterious *It* that takes control of the shooting process. This access to *It* is one of the main goals of all Asian martial arts.

Bruce Lee was acquainted with *It*. He experienced *It* in moments of inspired combat. In his teaching of Jeet Kune Do, he extolled the benefits of *It*, which he associated with the Taoist virtue of "no mind." In his book, *Tao of Jeet Kune Do*, he says, "The spirit is no doubt the controlling agent of our existence" (201). *No doubt.* That's the key— without doubt, being free of doubt. This is nothing other than *acting*. Lee goes on to declare, "Nirvana is to be consciously unconscious. That is its secret. The act is so direct and immediate that intellectual-ization finds no room to insert itself and cut the act to pieces" (201). He goes on to say, "When you are completely aware, there is no space for a conception, a scheme, 'the opponent and I'; there is complete abandonment" (204). This state of awareness is of the motion of that taking place in this. The "complete abandonment" is the act of giving up, of getting out of your own way. Lee says, "One can never be the master of his technical knowledge unless all his psychic hindrances are removed and he can keep his mind in a state of emptiness (fluid-ity), even purged of whatever technique he has obtained" (200).

Remember this above all: when all your psychic hindrances are removed, such as fear, doubt, conceptualization—including the conceptualization of yourself as the active agent distinct from what you are acting upon—and any sort of expectation, particularly of a specific outcome, you are open to the state of grace in which there is complete awareness. *You are anchoring that in this.* What does this do for your feeling of reality? *You can't go home again.* You will feel like a stranger in a strange land where you don't speak the native language or comprehend the local customs. You no longer have a dog in the fight. This Nirvana feeling—of being unconsciously conscious—will pervade your daily existence. Now it will be easy to live with uncertainty, and your initial doubt about the reality around you will settle into a comfortable indeterminacy. This will be as much that as that previously seemed to be removed from this.

Oddly, you will derive stability and comfort from your newfound fence-straddling once you abandon the idea that there is a fence to straddle. It will feel like being on a tour of an exotic locale. You will appreciate the scenery, be entertained by the locals, bargain at the shops, listen to the music, and enjoy the local festivals, *but you do not live there.* You are in a state of grace, and nothing in your surroundings has a claim upon you. You carry that with you wherever you go, making this appear like a billboard you pass on the highway on your way to another attraction.

Okay, so you're just passing through. What about the awful stuff, the loss, and the sorrow and pain? Let's go back to William Blatty and his departed son. Peter had vanished. But how could he vanish from a reality that had no substance in the first place? It's all about orientation. William Blatty was oriented toward the existence of a solid, physical reality. He was fooled. He accepted the terms of Peter's existence as a given, just as he accepted the terms of his own existence. Now he was not so sure. With less belief in the physical nature of reality, he suspected that reality might be multidimensional and in the broader sense, spiritual. Once he joined Bruce Lee in the conviction that "the spirit is no doubt the controlling agent of our existence," he had changed his orientation to one in which Peter could continue to exist as Peter but without Blatty having the means to experience him directly as in the holographic version of reality. Blatty claimed to have felt Peter's presence around him. Wishful thinking? Only if your orientation is still toward the physical.

Here's where your state of grace comes in—your anchoring of that in this. You are no longer tied to the physical delusion. Your sense of self is no longer dependent upon an independently existing physical reality to validate what you know. What you don't know is now beyond measure because it is **total awareness**. True knowing has no object. Through *acting*, Blatty can locate Peter. He must find the feeling of being in the state of grace, of being like the Zen archer about to loose his arrow at the target. He must carry this feeling around with him throughout every day. Then there is less interference between himself and Peter.

Developing this feeling is ordinarily pretty hard at first, but Blatty has the advantage of having been stunned by Peter's death. This automatically makes external reality less real. The operative dynamic here is continuing to experience a reality that has betrayed you. Peter is no longer a part of it, yet his father is still here. How can that be? Who's the goat in this transaction? Who or what set up this shell game? If these considerations don't begin to disoblige your sense of reality, nothing will. Let's look at how this plays out.

Say you have lost a beloved mate, companion, or cherished relative with whom you shared a major part of your life. This dearest aspect of your emotional matrix, the very heart of your most tender feelings, is here and now dead and gone.

Notice how incomprehensible that sounds. How can someone be dead and gone in the here and now? It's because here and now always implies a then and there. That's the trap. When you are still here and now, you are compelled to remember the one you have lost as having been "there" then. Where is "there" then if it is no longer here and now? Even if you return to a location here and now that you visited with your cherished one *then*, it isn't the same "there" that it used to be. That's because it doesn't *feel* the same anymore. Remember, reality is a feeling. If it is "there", here and now, it doesn't feel the same as it did then and there because it isn't in the same reality. Its universe of feeling has changed.

In our feelings, we mistakenly separate then and there from here and now. That's why our departed beloved seems dead and gone. He or she went from being here and now to being there then—and it is a terribly, terribly sad and unquestioning computation that our left brain, in its robotic insistence, must make. Something that was here now that is no longer evident must inevitably have been there then, and the emotionally crushing computation of that equation depends upon the presumed existence of time and space.

As sequential time and three-dimensional space are mental constructs only, your left-brain awareness is creating a hypothetical reality in which here and now and then and there have separate locations. Yet you create then and there out of the here and now. Here and now is the source. Then and there is a projection of here and now into a meta-dimension of itself that carries an emotional charge. By isolating then and there from here and now, you increase its emotional potential to become an autonomous morphic field. Now it assumes a reality equivalent to the morphic field of here and now. This is how you get blindsided by a bogus historicity into becoming enmeshed in an oppressively restricted version of reality. Nonetheless, then and there always remains a derivative of here and now.

You've known or encountered people who are said to be living in the past. They are stuck in the then-and-there half of a progressive equation that is endlessly self-referential. It begins again where it ends, postulating a then and there that can be validated only by further postulating a here and now that is its equivalent before the computation begins. It's like a temporal Mobius strip—a twist in the here and now, which can appear as a past version of itself in the present.

Okay, so we're into another dizzying paradox that generates *frisson* like hallelujahs at a tent revival. I've led you down this preposterous path (why is there never a post-posterous?) to call your attention to a reflexive behavior pattern we all adopt. It can embed us in this reality while forgetting that we are equally in that one. It is one of the master illusions of our sly illusionist of the here and now.

Here's where the rubber meets the road. You've lost a part of yourself when your beloved shifted frequencies. Now everything in your here and now painfully reminds you of when your beloved was there then. Notice that you are in a between-the-worlds state, but you have failed to appreciate the fact and withheld your acceptance of a two-world equivalency. You believe the here-and-now world to be real while the then-and-there world is an imaginative reconstruction in your memory. You are miserable in the here and now precisely because here and now is dramatically different, both physically and emotionally, from then and there. You can *feel* it. This feeling of difference disables a synchronous mode of communication between the worlds.

Carl Jung called synchronicity (a term he created) an "acausal connecting principle." Synchronicity is the more accurate term for what used to be called coincidence. The trouble with coincidence was that it was random, unpredictable, and essentially meaningless, even though its appearance was often strikingly apt. It just happened when you least expected it. It was a kind of luck or turn of fate. Synchronicity links things together in a meaningful way but without one thing actually being interconnected with another thing in a direct cause-and-effect relationship.

For example, you have been thinking of Joe, who moved away two years ago and hasn't been heard from since. Your phone rings. You answer it, and it is Joe calling. Two events occur, but one did not directly cause the other to happen within the accepted norms of consensus reality. The events are meaningfully connected after the fact of their occurrence. In consensus reality, a thought is not considered a tangible action that can of itself make something happen. Your thought of Joe did not cause him to place a call to you in this reality. Actually, his call was the result of an acausal connecting principle that functions outside of "normal" time and space in that reality. Synchronicity operates here and now independently of then and there because it is not subject to ordinary time constraints. Synchronicity transcends time. Remember this.

Synchronicity connects the world of here and now with the world of then and there. They exist simultaneously. It is only your *feeling* that they are different that disconnects them. So, moping around trying to re-experience then and there in the here and now is futile. You're only emphasizing a difference you are making up. Allow here and now to include then and there as another aspect of itself. *Then* you walk between the worlds. Your seemingly lost beloved is still connected to the here and now but in an acausal way (a meaningful relationship free of causation).

Recall that cause and effect is an illusion. Cause and effect is dualistic. Cause and effect operates in space-time. Cause and effect is predicated on the existence of a continuous reality. Don't feed the morphic field of then and there with your attention and emotion. Don't break the connection that synchronicity gives you. So long as you remain committed to walking between the worlds, then and there is always here and now, and so is the departed dear beloved who participates in both modalities in equal measure made inseparable by the synchronous connectivity of your love.

And what is the design behind these complex challenges, the false leads, the blind corners, the red herrings, and the dead ends? What is the pattern underneath the non sequiturs, the contradictions, and the paradoxes? Why is the game so damned convoluted? It seems that we are being appraised, assessed. How far has our individuation process progressed? How do we deal with the challenges of a seemingly physical reality? The constants are clear—aging, disease, loss, pain, and death. What do we do with those as inescapable conditions of existence even though they are part of the total illusion?

Even the holiest and most spiritually advanced go through these harrowing experiences. They each took their turn being perceived as dead and gone in the here and now. The Buddha died, Mohammed died, Jesus died (his resurrection is theologically irrelevant here; he went through the death experience), all the saints died, all the great spiritual masters died, all the gurus died. Everybody who has preceded you, enlightened or not, has died. They all went through a change of frequency into a different dimension.

"What have they got that I ain't got?" asked the cowardly lion. "Courage," chorused Dorothy, Scarecrow, and the Tin Man. This is one of the character traits I mentioned earlier, but it can include all the rest. Courage in the face of overwhelming loss, of terrifying prospects, can be derived from being in a state of grace, of having anchored that in this. There is a field of grace that can surround us, that can insulate us from the most crushing of human experiences. It's like a pure, fresh gust of mountain air, full of pine scent and rushing water, that lifts our hearts and draws us above ourselves so we say, "Yeah, this feels deadly, but I'm still consciously alive and just passing through. There's light down the pike at the end of the tunnel." After all, since the substance of the universe is consciousness, and consciousness processes information, then it is behavior that is of primary importance.

So the issue here, I feel, is how we stacked up during our sojourn in a holographic reality. Were we staunch? Did we learn from our mistakes? Did we see through the façade? Did we draw the best conclusions from what we went through? Did we appreciate the joke, and did we laugh? The answers to those questions will either open up further dimensions, more possibilities, and higher levels of awareness or leave us still struggling to make par. Yet no one is left behind. It's not in the cards. We are cherished and dearly loved by the universal consciousness out of which we continuously emerge. At some point, we will have learned enough to write, with full awareness, the scenario for ourselves. We will have learned to write, not a smarmy, stereotypical happy ending but a fascinating and entertaining ongoing.

Let's make it so.

Engage!

Bibliography

Blatty, William Peter. *Finding Peter: A True Story of the Hand of Providence and Life after Death.* Washington, DC: Regnery Publishing, 2015.

Castaneda, Carlos. *The Fire from Within.* New York: Washington Square Press, 1984.

Friedman, Norman. *Bridging Science and Spirit: Common Elements in David Bohm's Physics, The Perennial Philosophy, and Seth.* Woodbridge, NJ: The Woodbridge Group, 1993.

_____. *The Hidden Domain: Home of the Wave Function, Nature's Creative Source.* Woodbridge, NJ: The Woodbridge Group, 1997.

Gillabel, Dirk. "The Individuation Process." On the Internet at http://www.soul-guidance.com/houseofthesun/.

Goleman, Daniel. "Probing the Enigma of Multiple Personality." *The New York Times,* Science: June 28, 1988.

Lee, Bruce. *Tao of Jeet Kune Do.* Burbank, CA: Ohara Publications, 1975.

The Living Matrix—The New Science of Healing. DVD: The Living Matrix, LTD and Becker Massey LLC, 2009.

McTaggart, Lynne. *What Doctors Don't Tell You.* New York: Avon Books, 1998.

Roberts, Jane. *The "Unknown" Reality: A Seth Book Volume 1.* Englewood Cliffs, NJ: Prentice Hall, 1977.

Sanchez, Victor. *The Teaching of Don Carlos.* Rochester, VT: Bear & Company, 1995.

Scheinfeld, Robert. *Busting Loose from the Money Game: Mind-Blowing Strategies for Changing the Rules of a Game You Can't Win.* Hoboken, NJ: John Wiley & Sons, 2006.

Taylor, Jill Bolte. *My Stroke of Insight.* New York: Penguin, 2009.

Tolle, Eckhart. *A New Earth: Awakening to Your Life's Purpose.* New York: Penguin, 2005.

Watson, Lyall. *Gifts of Unknown Things.* Rochester, VT: Destiny Books, 1991.

Woit, Peter. "The Goldilocks Enigma: Why is the Universe Just Right for Life? by Paul Davies." Rev. of *The Goldilocks Enigma*, by Paul Davies. *New Humanist* 31 May 2007.

Printed in the United States
By Bookmasters